INTEGRATING NEW TECHNOLOGIES IN INTERNATIONAL BUSINESS

Opportunities and Challenges

INTEGRATING NEW TECHNOLOGIES IN INTERNATIONAL BUSINESS

Opportunities and Challenges

Edited by
Gurinder Singh, PhD
Alka Maurya, PhD
Richa Goel, PhD

First edition published 2022

Apple Academic Press Inc.
1265 Goldenrod Circle, NE,
Palm Bay, FL 32905 USA

4164 Lakeshore Road, Burlington,
ON, L7L 1A4 Canada

CRC Press
6000 Broken Sound Parkway NW,
Suite 300, Boca Raton, FL 33487-2742 USA

2 Park Square, Milton Park,
Abingdon, Oxon, OX14 4RN UK

© 2022 by Apple Academic Press, Inc.

Apple Academic Press exclusively co-publishes with CRC Press, an imprint of Taylor & Francis Group, LLC

Reasonable efforts have been made to publish reliable data and information, but the authors, editors, and publisher cannot assume responsibility for the validity of all materials or the consequences of their use. The authors, editors, and publishers have attempted to trace the copyright holders of all material reproduced in this publication and apologize to copyright holders if permission to publish in this form has not been obtained. If any copyright material has not been acknowledged, please write and let us know so we may rectify in any future reprint.

Except as permitted under U.S. Copyright Law, no part of this book may be reprinted, reproduced, transmitted, or utilized in any form by any electronic, mechanical, or other means, now known or hereafter invented, including photocopying, microfilming, and recording, or in any information storage or retrieval system, without written permission from the publishers.

For permission to photocopy or use material electronically from this work, access www.copyright.com or contact the Copyright Clearance Center, Inc. (CCC), 222 Rosewood Drive, Danvers, MA 01923, 978-750-8400. For works that are not available on CCC please contact mpkbookspermissions@tandf.co.uk

Trademark notice: Product or corporate names may be trademarks or registered trademarks and are used only for identification and explanation without intent to infringe.

Library and Archives Canada Cataloguing in Publication

Title: Integrating new technologies in international business : opportunities and challenges / edited by Gurinder Singh, PhD, Alka Maurya, PhD, Richa Goel, PhD.
Names: Singh, Gurinder, editor. | Maurya, Alka, editor. | Goel, Richa, 1980- editor.
Description: First edition. | Based on a conference organized by the Amity International Business School. | Includes bibliographical references and index.
Identifiers: Canadiana (print) 20210280050 | Canadiana (ebook) 20210280166 | ISBN 9781771889575 (hardcover) | ISBN 9781774639061 (softcover) | ISBN 9781003130352 (ebook)
Subjects: LCSH: International business enterprises—Technological innovations. | LCSH: Small business—Technological innovations.
Classification: LCC HD62.4 .I58 2022 | DDC 338/.064—dc23

Library of Congress Cataloging-in-Publication Data

Names: Singh, Gurinder, editor. | Maurya, Alka, editor. | Goel, Richa, 1980- editor.
Title: Integrating new technologies in international business : opportunities and challenges / edited by Gurinder Singh, PhD, Alka Maurya, PhD, Richa Goel, PhD.
Description: 1st Edition. | Palm Bay, FL : Apple Academic Press, 2022. | Includes bibliographical references and index. | Summary: "The international business sector has been completely revolutionized due to shifts in global economy, digitization, and the Internet. Integrating New Technologies in International Business: Opportunities and Challenges explores the rapid changes in technology that have affected businesses and social environments that are offering new challenges and opportunities for small to mid-size enterprises (SMEs) and start-ups. It highlights how businesses in emerging economies are implementing the new technological innovations to compete in the global market. The chapters in the volume provide valuable insight on many cutting-edge topics on new technology in the business environment and the new digital world, or Industry 4.0, including: Internet of Things (IoT) and customer relationship management Cross-cultural management Artificial intelligence Social media advertising Multichannel banking Digital payment technology Blockchain technology Augmented reality Eye-tracking analysis This book will be a valuable resource for business leaders and managers, industry professionals, business scholars, regulatory stakeholders, policymakers, faculty and students, and those who are interested in the current trends in the state of global digitization in industrial markets. The information provided here will help readers find the most appropriate approaches for taking advantage of these new technologies"-- Provided by publisher.
Identifiers: LCCN 2021036510 (print) | LCCN 2021036511 (ebook) | ISBN 9781771889575 (hardback) | ISBN 9781774639061 (paperback) | ISBN 9781003130352 (ebook)
Subjects: LCSH: International business enterprises--Technological innovations. | Small business--Technological innovations. | New business enterprises--Technological innovations. | Business enterprises--Computer networks.
Classification: LCC HD2755.5 .I68 2022 (print) | LCC HD2755.5 (ebook) | DDC 338.8/8--dc23
LC record available at https://lccn.loc.gov/2021036510
LC ebook record available at https://lccn.loc.gov/2021036511

ISBN: 978-1-77188-957-5 (hbk)
ISBN: 978-1-77463-906-1 (pbk)
ISBN: 978-1-00313-035-2 (ebk)

About the Editors

Gurinder Singh, PhD

Group Vice Chancellor, Amity Universities, Uttar Pradesh, India

Gurinder Singh, PhD, is Group Vice Chancellor of Amity Universities, Uttar Pradesh, India, as well as Director General of Amity International Business School. Professor Singh has extensive experience of more than 21 years (since 1998) in institutional building, teaching, consultancy, research, and industry. A renowned scholar and academician in international business, he holds a prestigious doctorate in the area along with a postgraduate degree from the Indian Institute of Foreign Trade, where he illustriously topped with seven merits. He holds the distinction of being the youngest Founder Pro-Vice-Chancellor of Amity University for two terms and the Founder Director General of Amity International Business School and the Founder CEO of the Association of International Business Schools, London. He has been instrumental in establishing various Amity campuses abroad, including in London, the USA, Singapore, Mauritius, and other parts of the world. He has spoken at various international forums, which include the prestigious Million Dollar Round Table Conference at Harvard Business School, Thunderbird Business School, New York University, University of Leeds, Loughborough Business School, Coventry Business School, Rennes Business School, Essex University, University of Berkeley, California State University, National University of Singapore, Massachusetts Institute of Technology, University of Massachusetts Lowell, Brunel Business School , University of Northampton, RMIT University, Swinburne University of Technology, Eduniversal Conference North-Umbria, and many more. He has received more than 25 international and national awards and has graced a host of talk shows on various TV channels. He is a mesmerizing orator and has the rare ability of touching the human soul. He is internationally well-known professor in the area of management. He is also known in the field of academics as an institution builder, writer, professor, distinguished academician, top-class trainer, international business expert, and the champion of the hearts of students.

Alka Maurya, PhD

*Head, International Business Department; Professor,
Amity International Business School (AIBS), Amity University, Noida, India*

Alka Maurya, PhD, is heading the International Business Department of Amity University and is working as Professor at Amity International Business School, Amity University, Noida, India. She is a computer science graduate and has done her master's in international business at the Indian Institute of Foreign Trade, New Delhi, India. Her area of specialization is international business, and she is engaged in teaching subjects including World Trade Organization, global business management, trade logistics and documentation, international trade risk management, and managing business in emerging markets to postgraduate students.

Dr. Maurya has over 22 years of experience in teaching, research, and consulting. During her tenure, she initiated and successfully launched several new projects. At AIBS, she launched undergraduate programs in international business and three research journals of the institutions, and she is engaged in curriculum development to integrate best business practices in the management curriculum in the area of international business.

During her last assignment with the Plastics Export Promotion Council (Plexconcil) as Deputy Director, she worked on a number of projects for promoting exports of plastic products from India. She conducted training programs on "Exporting Plastic Products from India" and "Foreign Trade Policy" during her tenure at Plexconcil. While working with the National Centre for Trade Information (NCTI), an UN-designated trade point in India, she prepared several market and product feasibility reports for the Ministry of Commerce, export promotion councils, trade promotion organizations, multinational corporation, and small and mid-size enterprises. She did one of the Ministry of Commerce project GS on the basis on which the GSP benefit was restored for commodities exported from India to the USA.

She has received training from the International Trade Commission on trade information sources on the Internet. She has authored various books, to name a few: *Covid-19: Strategies for New Normal* (Bloomsbury), *Integrating New Technologies in International Business: Opportunities and Challenges*, Apple Academic Press, *Disruption of Globalization* (Bloomsbury), *Globalization: Opportunities for Emerging Economies* (Ocean Publishing), and *Thrust Areas of India's Export, 1st Edition*, (Amity University Press). Her research works are published in reputed national and international research publications.

She has delivered talks at various international forums including the Association of International Business and Professional Management (AIBPM), Indonesia; conferences of AIB, India chapter; Institute of Company Secretaries of India, Delhi University; Alagappa University and many more.

She has prepared reports for the Services Export Promotion Council to ascertain the impact of GSP on tour operators in India.

Richa Goel, PhD

Assistant Professor of Economics and International Business,
Amity International Business School, Amity University, Noida, India

Richa Goel, PhD, is Assistant Professor of Economics and International Business at Amity International Business School, Amity University, Noida, India. She is a Gold Medalist in her Master of Economics with dual specialization accompanied with an MBA in HR. She also holds a Bachelor of Law. She has PhD in Management in area of Diversity Management. She has a journey of almost 20+ years in academic. She is consistently striving to create a challenging and engaging learning environment where students become life-long scholars and learners. Imparting lectures using different teaching strategies, she is an avid teacher, researcher, and mentor. She has to her credit numerous Research Papers in reputed national and international journals.

She is serving as a member of review committee for conferences and journals. She is handling many Scopus international peer reviewed journals for regular and special issue journals. She is acting as the special issue Editor of the *Journal of Sustainable Finance and Investment*, which is abstracted and indexed in the Chartered Association of Business Schools Academic Journal Guide (2018 edition) and Scopus. She has authored many books by Springer, IGI Global, Taylor and Francis, Nova, Bloombury, Ocean, etc. She is the Research Coordinator with the Amity International Business School and organizer for conducting international conferences. Her area of interest includes business restructuring, fusion of technology with international business, sustainability, human resource management and diversity management, Strategic Management. She is a member of different societies and professional bodies, including the World Economic Association, Bar Council, Human Resource Benchmarking Association, and many more.

Contents

Contributors .... *xi*
Abbreviations .... *xv*
Acknowledgments .... *xvii*
Preface .... *xix*

1. **Wealth and Welfare Business: A Sustainable Business Model**.................... 1
   Md. Mashiur Rahman

2. **An Impact of Adoption of the Internet of Things (IoT) on Customer Relationship Management (CRM) in the Banking Sector of India**.......... 13
   Parul Bajaj, Imran Anwar, and Imran Saleem

3. **Cross-Cultural Management: Opportunities and Challenges** 31
   Kavita Thapliyal and Mahendra Joshi

4. **Exploring the Relationship Between Personality and Work-Life Balance** 55
   Joshin Joseph

5. **Artificial Intelligence as Disruptive Technology: A Boon or a Bane in the Global Business Scenario** 71
   Seema Sahai and Saurav Lall

6. **Understanding Consumer Responses Towards Social Media Advertising and Purchase Intention Towards Luxury Products** 95
   Amna Ahmad and Bilal Mustafa Khan

7. **Multichannel Banking and Customer Experience: A Literature Review of Channels as a Moderator** 119
   Nidhi Verma and Mandeep Kaur

8. **Center of Main Interest (COMI): Perspectives and Challenges** 131
   Aditya Tomer, Sumitra Singh, and Abhishek Rohatgi

9. **Digitalizing Business Innovation** 149
   Ramamurthy Venkatesh

10. **Digital Payment: A Robust Face of Modern India** 173
    Priyanka Jingar, Ravindar Meena, and Sachin Gupta

11. **The Emerging Smart Supply Solutions in Fresh Fruits: India Matching the International Business Standards, New Formats, and New Technologies** 187

    Navita Mahajan and Frits Popma

12. **Conceptual Integration of International Marketing in India** 205

    Vikas Garg, Shalini Srivastav, Pooja Tiwari, and Sonam Rani

13. **Blockchain Technology and Its Utilization in Tracking Milk Process** ... 225

    Anita Venaik, Richa Goel, and Pooja Tiwari

14. **Impact of New Technology on Business** 233

    Monika Sharma and Vikas Garg

15. **Impact of Augmented Reality in Sales and Marketing** 253

    Lalit Kumar Sharma

16. **Theoretical Perspective of Role of Technology on Business Environment** 269

    Jasmine Mariappan, Chitra Krishnan, Karthick Shankaralingam, and Syed Mohd. Abbas

17. **Social Media Marketing and Purchase Behavior of Millennials: A Systematic Literature Review** 281

    Jitender Kumar, Sweta Dixit, Alka Maurya, and Ashish Gupta

18. **Eye-Tracking Analysis of Chosen Tourist Offers** 293

    Mariusz Barczak, Piotr Szymański, and Martin Zsarnoczky

Index *307*

Contributors

Syed Mohd. Abbas
RSM, Raymond Apparel Ltd., Mumbai, Maharashtra, India

Amna Ahmad
Research Scholar, Department of Administration, Aligarh Muslim University, Uttar Pradesh, India

Imran Anwar
Research Scholar, Aligarh Muslim University, Uttar Pradesh, India

Parul Bajaj
Post Doctoral Fellow, Aligarh Muslim University, Uttar Pradesh, India

Mariusz Barczak
University of Economy/Emotin sp. z o.o./Kodolanyi Janos University, Hungary

Sweta Dixit
Associate Professor, Sharda University, Noida, Uttar Pradesh, India

Vikas Garg
Associate Professor, Amity University, Greater Noida Campus, Uttar Pradesh–201308, India

Richa Goel
Assistant Professor, Amity International Business School, Amity University, Noida, Uttar Pradesh, India

Ashish Gupta
Assistant Professor, IIFT, New Delhi, India

Sachin Gupta
Assistant Professor, Department of Business Administration, Mohanlal Sukhadia University, Udaipur, Rajasthan, India

Priyanka Jingar
Research Scholar, Department of Business Administration, Mohanlal Sukhadia University, Udaipur, Rajasthan, India

Joshin Joseph
Jr. Accountant and ICWAI Trainee, Kerala Minerals and Metals Ltd. (A Govt. of Kerala Undertaking), Kerala, India

Mahendra Joshi
Associate Professor, Seidman College of Business, Grand Valley State University, Michigan, USA

Mandeep Kaur
Professor, University School of Financial Studies, GNDU, Amritsar, Punjab, India

Bilal Mustafa Khan
Department of Administration, Aligarh Muslim University, Uttar Pradesh, India

Chitra Krishnan
AIBS, Amity University, Noida, Uttar Pradesh, India

Jitender Kumar
Assistant Professor, Sharda University, Noida, Uttar Pradesh, India

Saurav Lall
Azure IoT, Microsoft Seattle, Washington, USA

Navita Mahajan
Associate Professor, Amity International Business School, Amity University, Noida, Uttar Pradesh, India

Jasmine Mariappan
Ibra College of Technology, Muscat, Oman

Alka Maurya
Professor, Amity University, Noida, Uttar Pradesh, India

Ravindar Meena
Research Scholar, Department of Business Administration, Mohanlal Sukhadia University, Udaipur, Rajasthan, India

Frits Popma
Managing Director, Popma Fruits Expertise, Netherlands

Md. Mashiur Rahman
Senior Executive Officer, Risk Management (CRM) Division, Bank Asia Ltd., Dhaka, Bangladesh

Sonam Rani
GL Bajaj Institute of Management and Technology, Greater Noida, Uttar Pradesh, India

Abhishek Rohatgi
Scholar, Amity Law School, Noida, Uttar Pradesh, India

Seema Sahai
Associate Professor, Amity International Business School, Amity University, Noida, Uttar Pradesh, India

Imran Saleem
Professor, Aligarh Muslim University, Uttar Pradesh, India

Karthick Shankaralingam
Freelancer, Oman, Muscat

Lalit Kumar Sharma
Associate Professor, Jaipuria Institute of Management, Ghaziabad, Uttar Pradesh, India

Monika Sharma
GL Bajaj Institute of Technology and Management, Greater Noida, Uttar Pradesh–201306, India,
E-mail: sharma.monika1785@gmail.com

Sumitra Singh
Assistant Professor, Amity Law School, Noida, Uttar Pradesh, India

Shalini Srivastav
Amity University, Uttar Pradesh, India

Piotr Szymański
University of Economy/Emotin sp. z o.o./Kodolanyi Janos University, Hungary

Kavita Thapliyal
Associate Professor and Chief Corporate Trainer (Business Communication and Soft Skills), Amity International Business School, Amity University, Noida, Uttar Pradesh, India

Pooja Tiwari
ABES Engineering College, Ghaziabad, Uttar Pradesh, India

Aditya Tomer
Additional Director/Joint Head Amity Law School, Noida, Uttar Pradesh, India

Anita Venaik
Professor, Amity Business School, Amity University, Noida, Uttar Pradesh, India

Ramamurthy Venkatesh
Research Scholar, Faculty of Management, Symbiosis International (Deemed University), Maharashtra, India

Nidhi Verma
Research Scholar, University School of Financial Studies, GNDU, Amritsar, Punjab, India

Martin Zsarnoczky
University of Economy/Emotin sp. z o.o./Kodolanyi Janos University, Hungary

Abbreviations

AGFI	adjusted goodness of fit
AI	artificial intelligence
AMA	American Marketing Association
AMPEE	analyze, model, process, execute, evaluate
AR	augmented reality
AVE	average variance explained
BHIM	Bharat interface for money
BM	business model
BPE	business performance evaluation
BPM	business process management
BS	business strategy
BVC	business value chain
CA	controlled atmosphere
CFA	confirmatory factor analysis
CFI	comparative fit index
CR	composite reliability
CRM	customer relationship management
CSR	corporate social responsibility
EIR	European insolvency regulations
GFI	goodness of fit
GLM	general linear modeling
HUL	Hindustan Unilever limited
ICT	information and communication technology
IDC	International Data Corporation
IDV	individualism vs. collectivism
IND	indulgence vs. restraint
IoT	internet of things
IRHM	International Human Resource Management
IT	information technology
KMO	Kaiser-Mayer-Olkin
LTO	long-term orientation vs. short-term orientation
MAS	masculinity vs. femininity
MNE	Multinational Enterprises
MSiT	Ministry of Sports and Tourism

OB	organizational behavior
PDI	power distance index
PMJDY	Pradhan Mantri Jan Dhan Yojana
POW	proof of work
PPI	pre-paid instruments
RMR	root mean square residual
RMSEA	root mean square error of approximation
SM	social media
TAM	technology acceptance model
TM	telemanagement
UAI	uncertainty avoidance
UNCITRAL	United Nations Commission on International Trade Law
UPI	unified payments interface
USP	unique selling proposition
VIDE	value, innovation, digital, and enterprise
VR	virtual reality
WLB	work-life balance
WWW	World Wide Web

Acknowledgments

We wish to express our sincere appreciation to those who have contributed to this book and supported us in one way or the other during this incredible journey.

First and foremost, we would like to thank with utmost devotion and humility "The Almighty Lord" for clearing all the obstacles and making it more an opportunity than a problem for this valuable work.

We would like to express our special thanks to Dr. Ashok K. Chauhan, Founder President Amity Group, for being the mentor, a guide, and a source of encouragement and support. He has been a living role model to us, taking up new challenges every day, tackling them with all his grit and determination, and always thriving to come out victorious. His vigor and hunger to perform in an adverse situation have inspired us to thrive for excellence and nothing less. It is rightly said that you cannot teach a person anything; you can only help him to find it within himself.

We would like to acknowledge the immense role of our families in enabling us to complete this project. We consider ourselves the luckiest in the world to have such a supportive families standing behind us with their love and support.

Many people played an essential role in developing this edition of the book, and we are deeply grateful to all of them.

—**Gurinder Singh, PhD**
Alka Maurya, PhD
Richa Goel, PhD

Preface

The international business sector has been completely revolutionized due to shifts in the global economy, digitization, and the Internet age. According to some research, it is the transition to the 3rd industrial revolution. Many believe that it is the beginning of the 4th industrial revolution. With the new wave of technological innovation, there has been the emergence of institutional settings, which has led to changes across the countries in the mechanism responsible for standardization. Such industrial transformation and rapid changes in technology, economics, and social environments create new opportunities for SMEs and start-ups. The new digital worlds, or industry 4.0, is changing the sectoral boundaries, deconstructing traditional industries, and stimulating the emergence of new sectors, thereby increasing flexibility impact upon labor market and employment practices.

India now has a massive economy with young aspiring global managers. Therefore, to preserve its international standing and further its economic prosperity in the 21st century, it should promote innovation by encouraging entrepreneurship and re-training of its workforce.

Moreover, the endeavor of this book is to create sky-scraping-quality, highly trained, and highly skilled graduates. This book is one such initiative, which is a collection of cutting-edge, high-quality articles backed by thorough and systematic research. This book includes scholarly chapters that focus on current business practices in all areas of management.

With these thoughts in mind, Amity International Business School organizatized a conference on the theme "Integrating New Technologies in International Business: Opportunities and Challenges." This compendium is an offshoot of the forum and highlights how emerging economies are integrating the new technologies and competing in the global market.

We specially thank the diverse and international review board members for their special review. Feedback and suggestions are sincerely invited from our readers to help further improve the quality.

We hope that this book enriches readers in contemporary areas of international business and helps them to understand the best practices from around the globe.

We wish readers an enjoyable reading experience!

CHAPTER 1

Wealth and Welfare Business: A Sustainable Business Model

MD. MASHIUR RAHMAN

Senior Executive Officer, Risk Management (CRM) Division, Bank Asia Ltd., Dhaka, Bangladesh

ABSTRACT

The aim of this study is to explore to what extent this business model (BM) can support a sustainable business strategy (BS). The course has been guided by the objectives to assess the present-day businesses' problems, the necessity of this new BM, a description of the model, its benefits, and the limitation of the proposed model. For the purpose of the study, a descriptive research has been focused on selected businesses, which were in operation at the time of the study and formed in a capitalistic way. Data have been collected mainly through in-depth interviews of entrepreneurs, employees, and customers. Secondary data have also been used by reviewing bulletins, newspapers, and articles from national and international journals. A sample has been selected by using the stratified random sampling method.

The findings of the study have shown that only money cannot give happiness and ensure sustainable development. Moreover, the economy is dominated by a system of interest-based borrowing. However, an interest-based borrowing system has several problems, like the certainty of expenditure by a business against the uncertainty of its profit. The purpose of a business should not be only profit maximization but also welfare. The employees should be treated as family members of a business enterprise. Corporate social responsibility (CSR) should be carried out in a structured way. The business house should maintain fair distribution and taxation policies. The study concludes that the proposed BM will help to develop a sustainable business structure.

1.1 INTRODUCTION

1.1.1 BACKGROUND OF THE STUDY

As per a report of BBC Bangla (2017), the wealth of only eight rich people of the world accounts for that of half of the world population. This shows there is massive inequality in the world as a few institutions and individuals mostly enjoy wealth. The report shows that inequality between the rich and the poor is increasing rapidly, and the inequality is far higher than imagined. Such a vast inequality will reduce the purchasing power of common people that will ultimately threaten businesses' sustainability. Besides, a report published in the daily Ittefaq (March 2018) mentioned that the happiest country in the world was Finland, and then after Norway comes Denmark, Iceland, and Switzerland. On the other hand, the world's highest displeased countries are Burundi, Central African Republic, South Sudan, Tanzania, and Yemen. The report also said that the USA was not among the top ten happiest countries earliest and now its position is 18th although most of the richest people are living in the USA. Under the above circumstances, to build a better world, people need wealth, but not at the cost of welfare, as welfare makes people happy, and happy people can make the economy sustainable. Welfare is lost when the environment and human attitude go against humanity. Moreover, the environment has multiple effects on sustainable business, which ensures economic growth, development, and happiness.

1.1.2 JUSTIFICATION OF THE STUDY

We find that the unequal distribution of wealth and dissatisfaction of the wealthy and the poor leads to economic uncertainty as unequal distribution of wealth reduces the majority's purchasing power, causing a recession in the economy. Dissatisfied people will concentrate less on work, which causes productivity loss and inefficiency of a labor force. As we discussed above, there is inequality and dissatisfaction in the business we do in the present-day world. It will threaten sustainability in the future. Therefore, it is essential to develop a business model (BM) which will be free from inequality and dissatisfaction. As such, a sustainable BM is essentially important for the benefit of an economy as well as humanity.

1.1.3 PURPOSE OF THE STUDY

The purpose of this study is to explore a BM for an enterprise that can ensure its sustainable growth.

1.1.4 OBJECTIVE OF THE STUDY

In general, the objective of the study is to identify the major factors that will help develop sustainable BM. Specifically, the study attempts to achieve the following specific objectives:

1. To determine the major principles for forming a business and characteristics of an entrepreneur for developing a sustainable BM.
2. To determine the independent variables that substantially affect the dependent variables, i.e., the sustainable BM.

1.2 LITERATURE REVIEW

Frank Boons et al. (2013) had conducted a study on sustainable innovation, BMs, and economic performance: an overview. The study has revealed that sustainable development needs radical and systemic innovations, which can be more effectively created and studied while building on BMs' concept. The sustainable concept gives firms a holistic framework of how to put in place sustainable innovations. The BM concept offers a connection of an individual firm with a greater production and consumption system in which it runs.

Tolkamp et al. (2018) also conducted a study on a user-centered sustainable BM design: the case of energy efficiency services in the Netherlands. The study finds that the firms hold interactions with their users in three stages: during designing, marketing, and the use-phase. For these separate phases, they also found an involvement loop consisting of four stages: design of involvement, facilitation of involvement, extraction of lessons learned, and finally, BM adaptation. Moreover, different types of involvement of users were found. They ranged from sending and receiving information to co-producing and co-innovating the BM. Involving users requires the scope for interaction on multiple components of the BM. It can lead to both incremental and radical BM innovation.

Bocken et al. (2015) conducted a study on value mapping for sustainable business thinking. The study finds that sustainable business necessitates companies to follow a systemic approach that considers three dimensions of sustainability-social, environmental, and economical in a way that creates shared value for all stakeholders, including the environment and society. The BM concept suggests a framework for system-level innovation for sustainability. It also provides the conceptual link with a firm's activities such as design, production, supply chains, partnership, and distribution channels. This study explores the use of value mapping for broader sustainable business thinking through reflection on its use in workshop settings.

Alexandre and Laymond (2016) carried out a study on the triple-layered BM canvas: A tool to design more sustainable BMs. The study reveals that the three layers make more evident how an organization creates multiple types of value, namely economic, environmental, and social value. Visually representing a BM through this canvas tool helps develop and communicate a more integrated view of a BM. It also leads to two new dynamics for analysis: horizontal coherence and vertical coherence.

Stefan Schaltegger (2016) conducted a study on BMs for Sustainability: A Co-Evolutionary Analysis of Sustainable Entrepreneurship, Innovation, and Transformation. The study finds that the BMs of sustainable niche market pioneers were identified in earlier research too, but little was known about the dynamics of the BMs in sustainable entrepreneurship processes targeting ecologically and socially beneficial niche models or sustainability upgradation of conventional mass-market players. The core processes of BM variation, selection, retention, and evolutionary pathways support structured analyses of dynamics involving BM innovation and the sustainability transformation of markets.

1.3 METHODOLOGY

1.3.1 INTRODUCTION

This study highlights the type of research design that has been used, the study population, the sample size, sampling procedures, data collection instruments, and procedure for data analysis.

1.3.2 RESEARCH DESIGN

The study contains diverse methods and tools that are relevant to achieve the desired research outcome. Accordingly, the research strategy has focused on the qualitative approach.

The purpose of this study's qualitative aspect is to seek information that can be generalized about the association between existing business and socially responsible sustainable BMs. The study is based on in-depth interviews and document analysis. The purpose of the qualitative study is to obtain a clear understanding of the business, employees, and customers as well as all other stakeholders, for which we have conducted detailed interviews of them. The study helps build a sustainable BM.

1.3.3 TARGET POPULATION AND SAMPLE

The study population involved entrepreneurs, customers, and employees of different locations. The target populations were selected purposively, and the survey was conducted in March 2019.

1.3.4 DATA COLLECTION INSTRUMENTS

Primary and secondary data were used; questionnaires were used to gather primary information from the field level. Open-ended questions were set and used. The open-ended questions were used to encourage the interviewees to provide extensive answers. Their comments and suggestions were also included. Secondary data was obtained through the website and external sources. Expert opinions were also taken from Senior Entrepreneurs, University Professors, and Senior Bankers.

1.3.5 METHOD OF DATA ANALYSIS

Qualitative data were collected, and a BM was formed comprising the five stages, such as:
1. Present business and their problem;
2. Necessity of new business model;

3. Description of the new business model;
4. Benefits of a new model, and finally; and
5. Limitation of the proposed model.

Finally, the BM was presented using graphics and a description of graphs.

1.4 FINDINGS/BUSINESS MODEL (BM)

1.4.1 DEMOGRAPHIC CHARACTERISTICS

According to the results obtained from the data, out of a total of 100 expert respondents, 20 were Entrepreneurs, 30 employees, and 50 customers. Of them, 90.50% were males and 9.50% females. Among the respondents, the age of 46 was above 40, while 28 were aged between 36 and 40. A total of 26 were aged between 31 and 35.

1.4.2 WEALTH AND WELFARE BUSINESS MODEL (BM)

1. **Present Business and Their Problem:** From different reports, we find that wealth is unequally distributed, wealthy persons are not happy, and the poor are struggling to meet their basic needs. In this connection, the then head of the IMF (Daily Ittefaq, June 2018) thought a black hole in the economic sky. Some models are wrongly described, which drives humans in the wrong direction with the worst results. It is like the capital-structured theory of M. Millar, where it is mentioned that due to taxation, the interest-based capital structure is better than equity-based capital structure, but interest is an expenditure which is certain while profit is uncertain. This certain payment of interest can make an enterprise bankrupt, as was the recession in the USA in 2007–2008 due to over-financing. Moreover, there is no structure for expenditure under corporate social responsibility (CSR). On the other hand, tax mobilization is a kind of robbery that deprives a section of people to the benefit of others. As such, sustainable BM is essentially important.
2. **Necessity of Wealth and Welfare Business Model (BM):** In view of the above, we find that society revolves around wealth, but that is not sustainable, and a recession may occur any day. As such, the then head of IMF (Daily Ittefaq, June 2018) was very much concerned.

This scenario underlines the necessity of forming a new BM for society's well-being, as happiness and sustainability depend on wealth and social welfare.

3. **Description of Wealth and Welfare Business Model (BM):** The wealth and welfare BM (a sustainable BM) have been developed with seven principles and seventeen characteristics of an entrepreneur (Figure 1.1). The principles are:

   i. The entrepreneur will be skilled and honest, and the business will be interest-free.
   ii. The business will be run, keeping in mind the welfare of human beings.
   iii. Single or more than a single entrepreneur can be there, and they will sit on the Board of Directors and take the business responsibility. The responsibility can be limited by capital. The profit can be taken at last.
   iv. There will be a special kind of general investor(s) who will not take a loss, but they can get profit at the rate of zero (0) to infinitive as per contract. This type of investor cannot sit in the meeting of the board of directors.
   v. Laborers will get comparatively high wages in consideration of business income.
   vi. A portion of the income will be distributed among the poor as Zakat or CSR.
   vii. In the wealth and welfare business, the tax will be provided to the government for nation-building.

The characteristics of an entrepreneur should be as follows:

- He should be a believer;
- He should avoid the business of products and services that go against humanity;
- He must nor work based on assumption;
- Maintains the commitment;
- Keeps goods of others safe;
- Pays Zakat personally;
- He is cool and realistic;
- He does not make bad comments against other religions;
- Doesn't lie;
- Not arrogant and talks with decency;
- Respects the local and international laws;

- Knowledgeable enough about his/her business;
- Consults with specialists before starting any business;
- Appreciates better work by others;
- Maintains good relations with parents;
- Maintains good relations with relatives and neighbors;
- Doesn't abuse others sexually.

Notably, if the entrepreneur is not a good human being, he cannot be a good entrepreneur.

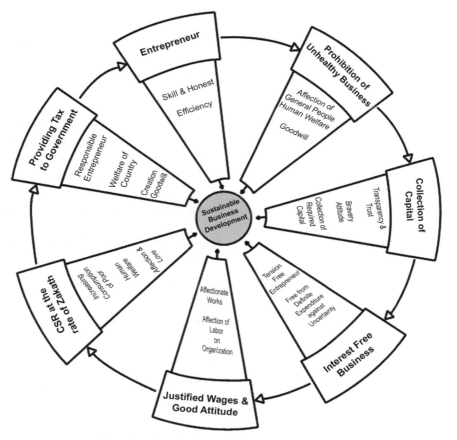

FIGURE 1.1 Wealth and welfare business model.

In the BM, there are seven independent variables which are listed below: (a) entrepreneur, (b) prohibition of unhealthy business, (c) collection of capital, (d) interest-free business, (e) justified wages and a good attitude, (f) CSR at the rate of Zakat, and (g) providing taxes to the government. All of the independent variables independently and accumulatively affect the dependent variables, i.e., sustainable business development. The independent variables also affect the dependent variables with some of its characteristics, which are lying between independent and dependent variables. Description of all independent variables and how they play their role in sustainable business development are as follows:

a. **Entrepreneur:** An entrepreneur is a person who takes the first step towards doing a business and takes the risk. The business will develop and be secure, depending on the entrepreneur's honesty and skill, which will ultimately help create sustainable business development efficiency.

b. **Prohibition of Unhealthy Business:** A man involved with wealth and welfare business must not engage himself in any unhealthy business. He employs himself for humanity's welfare that will earn him the appreciation of the general people and help in sustainable business development.

c. **Collection of Capital:** Capital is the blood circulation of business. Both excess and shortfall of capital are harmful to a business. A wealthy and welfare businessman will follow the golden mean between this. For the purpose of the collection of capital, transparency and trust have to be maintained. It will help garner courage and thus collect the required capital.

d. **Interest-Free Business:** Interest means anything certainly in excess against lending of money or assets or lately charges of penalty against credit sales. Interest makes not only poor people poorer but also makes rich people poorer. As interest is the expense, which is certain while there is the uncertainty that the business will earn a profit. The interest-free business will help an entrepreneur remain tension-free and growth in business.

e. **Justified Wages and Good Attitude:** Labor is the vital driving force of any business enterprise. The labor force's willingness to work and unwillingness to do so are not equal ever. Logical and good behavior makes workers more loyal and feels their attachment to the enterprise.

f. **CSR at the Rate of Zakat:** For the purpose of shared prosperity, we should pay a portion of income to the poor without expecting any direct benefit. This expenditure will increase the consumption of the poor, ultimately pushing up the sales of the business enterprise. So far, Zakat should be the barometer for expenditure under CSR as there is no alternative expenditure rate for shared prosperity.

Notably, Zakat means donating some of the personal wealth to those in need. Zakat purifies the human spirit, brings a man closer to Creator, and ultimately helps to increase wealth. The Zakat is not payable for all. A certain amount of wealth, which we call nisab is payable. The nisab value is directly related to the current values of gold and silver. In the Holy Quran, the nisab value is described as 612.36 grams of silver or 87.48 grams of gold. Originally, these values of silver and gold were equal, but in the current days, they have become different amounts in value, with the silver nisab being significantly lower than its gold counterpart. Nisab value can be calculated at any given time by converting these weights into the local currency.

Zakat donation should amount to 2.5% of an individual annual total wealth accumulated over the year. For example, if total assets (after any debts owed) amounts to $10,000, then $250 is required to be paid as Zakat.

g. **Providing Tax to the Government:** An entrepreneur should pay a fair amount of tax for getting benefits of security, infrastructure, etc., from the government. Through taxation, the state gets money and can expend the money for the welfare of the people. Taxation increases the goodwill of a business enterprise.

All of the seven steps will accumulatively affect the sustainable business development model, i.e., wealth and welfare BM.

4. **Benefits of Wealth and Welfare Business:** Through the application of wealth and welfare BM, both the rich and the poor will be benefitted, and as such, the rich will be richer, and the poor also will be rich and powerful as CSR is considered as part of Zakat. Besides, interest has been prohibited. Importance has been given to personality with a good attitude. Honesty and efficiency, as well as confidence and bravery, also have importance in this BM. As such, the wealth and welfare BM will help build a sustainable business community and, thus, a happy society.

5. **Limitation of the Wealth and Welfare Business Model (BM):** The said BM would be helpful for only the people who want to be happy

and build a wealthy society. However, for those who want to acquire wealth by any means, the model will not give any result.

1.5 CONCLUSION

The study has revealed that for building a sustainable business society, the Wealth and Welfare BM is accurately perfect for humans to follow. However, in-human beings will not get any guidance from the BM. Here the seven principles to run a business and the seventeen characteristics a good entrepreneur needs are required to be followed strictly. These require an entrepreneur to be a good human being first. Lastly, the BM is given in five steps. The entire process is scientific and so helpful for the formation of a sustainable business society.

KEYWORDS

- **corporate social responsibility**
- **interest-free business**
- **sustainable business society**
- **sustainable model**
- **taxation**
- **wealth and welfare business model**

REFERENCES

Alexandre, J., & Laymond, L. P., (2016). The triple layered business model canvas: A tool to design more sustainable business models. *Journal of Cleaner Production.*

Bocken, N. M. P., et al., (2015). Value mapping for sustainable business thinking. *Journal of Industrial and Production Engineering, 32.*

Frank, B., et al., (2013). Sustainable innovation, business models, and economic performance: An overview. *Journal of Cleaner Production, 45.*

Lagarde, C., (2018). *Daily Ittefaq.* [Online] Available at: https://archive1.ittefaq.com.bd/trade/2018/06/13/160283.html (accessed on 21 October 2020).

Oxfam, (2017). *Prhothom Alo.* [Online] Available at: https://www.prothomalo.com (accessed on 21 October 2020).

Stefan, S., Lüdeke-Freund, F., & Erik, G. H., (2016). Business models for sustainability: A co-evolutionary analysis of sustainable entrepreneurship, innovation, and transformation. *Organization and Environment, 29*(3).

Tolkamp, J., et al., (2018). User-centered sustainable business model design: The case of energy efficiency services in the Netherlands. *Journal of Cleaner Production, 182*.

United Nations, (2018). *Ittefaq.* [Online] Available at: https://archive1.ittefaq.com.bd/national/2018/03/16/150656.html (accessed on 21 October 2020).

CHAPTER 2

An Impact of Adoption of the Internet of Things (IoT) on Customer Relationship Management (CRM) in the Banking Sector of India

PARUL BAJAJ,[1] IMRAN ANWAR,[2] and IMRAN SALEEM[3]

[1]Post Doctoral Fellow, Aligarh Muslim University, Uttar Pradesh, India

[2]Research Scholar, Aligarh Muslim University, Uttar Pradesh, India

[3]Professor, Aligarh Muslim University, Uttar Pradesh, India

ABSTRACT

Technology nowadays has become an important part of our everyday life. Innovations and up-gradation in technology help organizations and businesses to survive the massive global competition. In this paper, the impact of the Internet of things on Customer relationship management (CRM) has been discussed. The paper starts with a purpose to find out the impact of Cost, Convenience, Status, and Privacy on the adoption of the Internet of things (IoT) in banks. Later on, the impact of IoT adoption on CRM has also been checked. The CRM services have been defined with the help of three most important terms, responsible for the smooth CRM in banks, Responsiveness, Satisfaction, and Assurance. The responses of 366 educated customers from public and private banks have been taken for the study through questionnaires comprising 29 items. The Confirmatory Factor Analysis (CFA) and the Structural Equation Modeling (SEM) have been used to test the model fitness, reliability, and validity of the data and to test the hypotheses, respectively. The results reveal that not only Cost and Privacy have a significant impact on the adoption of IoT, but the Adoption of IoT also has a significant impact on CRM of banks. The present paper attempts to help the policymakers and customers who are inclined to know about the impact of newer technologies on the application of services in a better and easy way so that these services can be easily adaptable in the public interest.

2.1 INTRODUCTION

As we know in the course of recent decades, the world has been an unmatched evolution of information technology (IT), which has influenced our life a lot. Almost every single industrial segment has been affected, especially the services sector, like insurance, banks, etc. The banking sector is experiencing fast technological innovation, changes, and advancement (Al-Fahim, 2013). Now, internet innovations are connected very closely to the daily life of human beings covering different fields from the World Wide Web (WWW) to e-mail to e-learning portal to search engine to social networking to different online payments portal. It is rational to expect that another age of Internet will show up in the near future because of the advancement of internet technology and the development of the Internet of things (IoT) (Tsai et al., 2014). Most probably, more uses of IoT will be discovered, and people will be more inclined to use these services.

IoT can be defined as a concept that considers as the unavoidable presence of an assortment of things or objects through which people can be able to interact with each other and connected with other things or objects to create new applications or services and reach respective desired goals through wireless and wired connections and unique addressing schemes (Thibaut, 2014). The IoT makes people and objects able to be connected with Anything, Anyone, Anytime, and Anyplace, basically using any connected device and any service. No doubt, these devices will generate big data to be analyzed for knowledge extraction that makes organizations capable of offering the right products to the right customers (Perera et al., 2015).

With the advancement of branchless banking services through multiple communication channels, have made it possible to offer a new kind of value-added services for consumers, in the banking sector as well as other financial services sector. With the progressive utilization of smartphones and other wireless devices, such as wearable and sensors, has made the IoT as a new frontier for CRM to improve customer experience, intelligent development in electronic banking. In spite of providing good quality service via online and mobile banking systems, the quality of banking services can be made better by adopting IoT technologies (Jerald et al., 2015). Therefore, the banking sector has been trying to utilize the potential of IT for providing better services to the customers (Al-Hajri, 2008).

New technologies of collecting data have affected both factors; companies and customers. Now banks regularly inform the customers through various communication mediums. With the help of IoT, banks can change their way of marketing and the possibilities of gathering customer data for better CRM (Awasthi, 2012). In this way, CRM enables any firm to better acquire, manage, serve, and get the maximum profit from its customers (Customer relationship management (CRM), 2009). In this way, by adopting IoT, organizations will be able to interact with the customer on a real-time basis because of the customer data that will ultimately help the organization to maintain a good relationship with its customers. Therefore, the present study points out an important empirical belief that banks are competent enough to introduce important CRM changes regarding IoT. The purpose of this study is to find the impact of IoT on CRM after testing the impact of cost, convenience, status, and privacy on the adoption of IoT. Further, it checks the impact of these influencing factors (if any) in the case of customers from different banks.

2.2 REVIEW OF LITERATURE

In the present study, four customer's adoption of IoT influencing factors have been used named as cost, convenience, status, and privacy as well as their impact on CRM in banks has also been checked (if any). These attributes have been chosen on the basis of the literature review of the related studies done in India or abroad. These six attributes have been finalized as per the results of previous studies as well as repeated citations of various authors. Furthermore, very few studies, in general, concluded the impact of IoT on CRM. The hypotheses were framed in support of the situation that occurred after having an idea of the previous result of various literatures from the area.

2.2.1 FACTORS AFFECTING ADOPTION OF IOT

IoT has been defined by Cisco as smart objects, linked to the Internet and enable the exchange of accessible data and attract customers more securely (Cisco, 2015). Previous studies showed many factors that influence the adoption of IoT technique. Some factors have often shown a significant relationship with the adoption of IoT, such as perceived ease of use, convenience, security, trust, and awareness (Sicari et al., 2015; Farooq and Waseem, 2015; Jing et al., 2014; Elkhodr et al., 2013; Hsu et al., 2013). These factors are studied in some developing countries such as Mauritius (Li and Wang, 2013); in Oman (Al-Hajri, 2008); in Jordan (Al-Majali and Mat, 2011); in Malaysia (Suki, 2010) and in Tunisia (Nasri, 2011) demonstrated a significant relationship with adoption of IoT. It is used to be claimed to IoT as essential for a firm's innovation, success, and adaptation, especially for firms with a high amount of connectivity, network, and data (Janes, Suoranta, and Rowley, 2013; Yu, Nguyen, and Chaen, 2016).

2.2.2 COST

"The amounts of economic outlay that must be sacrificed in order to use IoT" (Kim et al., 2007). Chang and Wildt (1994) found that the cost of the services provided is directly having an influence on the rate of the adoption of IoT. If the cost is more than its utility, the customer takes less interest to adopt the service considering it expensive.

Cheong and Park (2005) also found IoT cost as the difference between payments made for using the services and benefits of using IoT. Kin and Shin (2015) also approved that there is an important relationship between IoT cost and its adoption. Cost is widely accepted as one of the major factors for influencing the buying decision of the customers. Many researchers confirmed a significant role in the cost of adopting IoT. Many studies revealed about the price sensitiveness regarding the adoption of the IoT and the relationship between the price of service and frequency of usage; for example, Cho and Sagynov (2015) also found that IoT users are more unstable towards the cost because of the availability of different prices for different services. Bajaj et al. (2019) also supported in her research results that there is a significant impact of cost on the adoption of IoT-enabled services. Thus, the first hypothesis is framed as follows:

> ➢ *H1: There is a significant impact on the cost of service on the adoption of the IoT.*

2.2.3 CONVENIENCES

The meaning of the term "convenience" is differently taken from time to time from the anatomy of a particular product with a special emphasis on time saving and easy to use (Cho and Sagnov). In the opinion of Tatnall and Davey (2017), convenience is a special factor that affects the adoption of the

Internet, and it is also proved by many researches. Basically, the concept of convenience indicates time-saving, get rid of overcrowded markets as well as parking issues, 24 hours access through online shopping. While IoT provides real-time information by which the service providers are able to offer the best desirable services as per customer's need. Gao and Bai (2014) also explained that easy-to-use is the main feature that attracts customers to use IoT. This result is supported by the technology acceptance model (TAM) that is easy to use as a necessary factor to attract customers and to use it as modern technology (Kuo and Yen, 2009; Lee et al., 2012; Venkatesh et al., 2012). In addition, Pikkarainen et al. (2004) also discussed that Internet-connected services are some of the cheapest and effective marketing channels as it is very convincing to use it. As well as Maditinos, et al. (2013) confirmed that by using online banking services, a bank could improve its services to get better access to its customers. Likewise, the study will try to check through its second hypothesis:

> ➢ *H2: There is a significant impact of Convenience in use on the adoption of the Internet of things*

2.2.4 STATUS

According to Gao and Bai (2014), the social aspect cannot be ignored in the case of using new innovations and technology. IoT is one of the new technology that emerged among the customers, especially in India; comparatively, it is a new phenomenon. Basically, the status factor has an impact on customer's behavior in the development stage, Hsu and Lu (2004). According to Venkatesh et al. (2012), the social context can be explained as a customer's perception of how other people in society gave importance to their behavior. The researchers mentioned the status influence as a person's belief that researchers should use new technology to find acceptance for other important people in society. Therefore, it can be concluded that the influence of family, friends, and peer groups is an essential factor in the adoption of IoT. Devis et al. (1989); Chong et al. (2012) also found an important role of status or social influence in adopting mobile commerce or any IT usage and acceptance. In the same quest, Al-Momani et al. (2016) concluded that the social status factor has a significant influence on customer's intention of adopting IoT. Therefore, the researchers frame the third hypothesis for the study is as follows:

> ➢ *H3: There is a significant impact of Social Status on the adoption of the Internet of things.*

2.2.5 PRIVACY AND SAFETY

As it is possible to interconnect all the devices and gather the information through the Internet, information from their day-to-day management of activities is easily accessible. Different gadgets like smartwatches, Google glass, or smart health kits, and so many are the sources for the information of high importance from the consumer. In this case, it is mandatory to improve the confidence level of the customer, Talari et al. (2017). In the era of IoT, the privacy and safety issue attracts significant attention from various researchers and other authorities, etc., even the government used to try to control the safety and privacy issue using the access of the data of the customer Decker and Stummer (2017).

No doubt IoT data is much more at risk rather than web data because it collects real-time information of the users, so privacy concern is an important concern in the adoption of IoT. As per the survey of the present study, around 60% of IoT users have the basic information regarding safety concerns like data collection of user's information. The survey also concluded that 87% of users are keen to have knowledge about personal information collection (TRUSTe, 2014).

Medaglia and Serbanati (2010) also found that security and privacy concern is an important issue to handle for user-oriented applications. As well as Cho (2004) identified that new technology offers many benefits but with some risk in which some risk may be high. Similarly, Luo et al. (2010) and Kim and Lennon (2013) explained that that the risk is also associated with the usage of services related to IoT. In the era of technology, the customers are very much concerned about the personal, especially the finance-related data (Weber, 2010). Hsu and Lin (2018) also supported that the weak handling of private data may negatively affect the adoption of IoT. Hence the fourth hypothesis for the study is as follows:

> ➢ H4: *There is a significant impact of Privacy and safety concerns on the adoption of the Internet of things.*

2.2.6 IOT AND CRM

Ou and Sia (2003) stated that the use of technology for the real-time information of the customer enables the organization to track their needs. The producer got a lot of data regarding the customers' consumption; therefore, they can have better access to their future, present as well as to past customers.

With the help of new online tools, the way to interact with the customers has been changed radically. A new age CRM has emerged (Parihar, 2012). Bompolis and Boutsouki (2014) supported that social media has become a new source for customer information and interaction with them. Bajaj et al. (2019) discussed that IoT had become a new frontier for better CRM. CRM can be explaining by its different dimensions for service excellence. Therefore, the present study will try to check whether:

- H5: *There is a significant impact on the adoption of IoT on the CRM dimension of employees' responsiveness in banks.*
- H6: *There is a significant impact of the adoption of IoT on the CRM dimension of Customer satisfaction in banks.*
- H7: *There is a significant impact of IoT on the CRM dimension of assurance in banks.*

2.3 METHODOLOGY

In this manuscript, authors have attempted to measure the impact of cost, convenience, status, and privacy, respectively, on the adoption of IoT, along with checking the impact of the adoption of IoT on CRM in the public and private banks. The adoption of IoT was formed by taking the variables; responsiveness, satisfaction, and assurance through the application of second-order CFA. Cross-sectional data from customers of three public and three private banks were collected using a self-developed questionnaire. Convenience sampling method (Anwar and Saleem, 2018, 2019; Krueger et al., 2000; Fayolle and Gailly, 2015; Linan and Chen, 2009) was used to collect data outside the bank branches in Aligarh, Uttar Pradesh, India. The target population for the study was customers of three public banks (SBI, PNB, and BoB) and three private banks (ICICI, HDFC, and AXIS). Out of total 450 questionnaires (75 each bank) distributed, 397 were retrieved and processed further for the data screening process. Further, in the data screening process, 35 questionnaires were removed, thus forming a final sample of 366 responses. Out of 366 respondents, 189 were found to be male, while 177 were female respondents aging between 18 to 65 years.

2.3.1 *DEVELOPMENT OF DATA COLLECTION INSTRUMENTS*

The survey has been done with the help of a self-structured questionnaire that contains four variables that are the influencing factors for the adoption of IoT that are cost, convenience, status, and privacy. The variables have been chosen from the literature available from in or outside India. The questions were farmed, taking in concern regarding the Indian context. To check the impact of IoT on CRM, three dimensions have been taken from the SERVQUAL (service quality) model (Persuraman et al., 1985; Bajaj et al., 2014), i.e., responsiveness, satisfaction, and assurance. The questionnaire has 29 questions under seven variables.

2.3.2 DATA SCREENING

The data screening process was carried out to ensure that the dataset, to be used in the present study, is cleaned and appropriate for further statistical analysis. Data were collected on a five-point Likert scale using convenience sampling method (Anwar and Saleem, 2018, 2019; Krueger et al., 2000; Fayolle and Gailly, 2015; Linan and Chen, 2009) from customers of three public (SBI, PNB, and BOB) and three private banks (ICICI, HDFC, and AXIS) respectively. Seventy-five questionnaires were distributed to the customers of these six banks directly in physical form outside of various branches in Aligarh city of Uttar Pradesh, India. Out of total 450 distributed questionnaires, total 397 questionnaires were filled and returned by respondents. Later, 19 questionnaires were found, having missing values; hence was removed. Although, median replacement method could be used to impute data for 19 questionnaires containing missing values (Kline, 1998; Cohen and Cohen, 1983), but authors found it suitable to remove those responses considering that enough size of the sample was available for the study. Furthermore, nine questionnaires were found to have unengaged responses; hence they were also eliminated from the total sample size. Outliers were also checked by authors using Cook's distance method, and it was found that statistics for all the responses were below the recommended threshold of 1 except three responses; thus, they were removed from the dataset, henceforth forming a final sample size of 366 responses (Table 2.1). The questionnaire used for the current study comprises 29 items; hence sample of 366 responses is justified for using structural equation modeling in AMOS since it is advisable to have at least 10 responses for each item used in the study (Kline, 1998). The normality of the data is also another assumption for using AMOS, thus ensuring the normality, skewness, and kurtosis measures were used. Statistics for skewness and kurtosis for all the constructs were found to be between the range of -2 and $+2$ hence confirming that the data is free from non-normality issues (Kline, 1998).

TABLE 2.1 Data Sample Synthesis

Public Bank		Private Bank		Gender	
SBI	62	Axis	59	189	Male
PNB	58	ICICI	63	177	Female
BOB	64	HDFC	60	**366**	**Total**
Total	**184**	**Total**	**182**		

Despite taking measures of reducing common method bias (psychological separation) into consideration while collecting the data, Harman's one-factor test (Podsakoff and Organ, 1986) was performed to ensure that the majority of variance (50%) is not explained by one single factor. It was found that only 32.112% (see Table 2.2) variance could be predicted by one factor, which is well below the suggested limit of 50%. Therefore, the present study seems not to be affected by any concern of common method bias.

TABLE 2.2 Total Variance Explained (Harman's Single Factor Test)

Component	Extraction Sums of Squared Loadings		
	Total	Percentage of Variance (%)	Cumulative (%)
1	10.109	32.112%	32.112%

2.4 RESULTS

2.4.1 MEASUREMENT MODEL: FIT INDICES, RELIABILITY, AND VALIDITY

Confirmatory factor analysis (CFA) was performed to have a qualitative assessment of the model since it ensures the validity and reliability of the constructs along with the model fit of the data which is being used in the study (Henseler et al., 2009). Since the study is focused on measuring the impact of cost, convenience, status, and privacy on IoT adoption at the first stage and at the second stage, the impact of the adoption of IoT on CRM in the banks, variables; cost, convenience, status, and privacy along with the adoption of IoT were diagramed into AMOS taking their respective observed variables using zero-order CFA while CRM was measured through responsiveness, satisfaction, and assurance using second-order CFA. Initially, first-order CFA was run taking all the constructs and fit indices were found to be adequate with the following values; CMIN/DF = 2.291,

GFI = 0.901, AGFI = 0.859, NFI =.891, CFI = 0.898, RMSEA = 0.061. Individual loadings of each item were also observed and found that all the items are converging to their respective factors with adequate coefficient weights, thus averaging near 0.70 except for Cost and Privacy, which were left with only two items having average loading near 0.70. Worthington and Whittaker (2006) suggested retaining the factor with only two items if average item loading is 0.70 or above while relatively less correlated with other factors (referring to discriminant validity), which is found to be the case here regarding Cost and Privacy thus retained into the study.

Second-order CFA was performed having the second-order reflective measurement for CRM, taking responsiveness, satisfaction, and assurance as factors of measurement. Responsiveness, satisfaction, and assurance were found to be converging into CRM with the loadings well above 0.70, which is desirable for second-order CFA. Thus forming the final CFA model, following model fit indices was achieved: CMIN/DF = 2.045, GFI = 0.914, AGFI = 0.886, NFI = 0.908 CFI = 0.921, RMSEA = 0.055. Enough improvement from the zero-order CFA model was witnessed in final order CFA in terms of model fit indices, and all the fit indices were found adequate to be called a good-fit model (Table 2.3).

TABLE 2.3 CFA Model Fit Indices

Model	CMIN/DF	GFI	AGFI	NFI	CFI	RMSEA
Recommended Value	Acceptable 1–4	≥0.90	≥0.85	≥.90	≥0.90	<0.07
References	Wheaton et al. (1977)	Shevlin and Miles (1998)	Shevlin and Miles (1998)	Hu and Bentler (1999)	Hu and Bentler (1999)	Mac Callum et al. (1996)
Model Fit Indices (First-Order CFA)						
Study Model (First-order)	2.291	0.901	0.859	0.891	0.898	0.061
Model Fit Indices (Second-Order CFA)						
Study Model (Second-order)	2.045	0.914	0.886	0.908	0.921	0.055

2.4.2 RELIABILITY AND VALIDITY

The reliability and validity of the constructs were also assessed to ensure whether observed items are having more of the convergence into their respective factors than factors correlations with other factors. Construct reliability was measured using composite reliability (CR) and Cronbach's Alpha Reliability. A score of 0.60 or above for CR is termed as acceptable (Bagozzi and Yi, 1988), while for Alpha reliability, a score of 0.70 or above is considered as desirable (Hair et al., 1998). For the current study, values for CR have been found ranging from 0.688 to 0.922, while values for Alpha reliability have been found in the range of 0.683–0.906. Furthermore, convergent and divergent validity were also measured using James Gaskin's statistical tool package. The average variance explained (AVE) was found to be above 0.50 for all the constructs except for cost. Hair et al. (1998) suggested the acceptable threshold for AVE to be at or above 0.50, which is met here in this study except for cost (Table 2.4).

TABLE 2.4 CFA Loadings, Cronbach's Alpha, CR, AVE

Construct Name	Variable Name	No. of Indicators	Avg. CFA Loading	Alpha	CR	AVE
Cost		2	0.692	0.683	0.688	0.480
Adoption		4	0.794	0.864	0.875	0.638
Convenience		4	0.733	0.821	0.824	0.541
Status		3	0.718	0.761	0.762	0.517
Privacy		2	0.708	0.667	0.669	0.502
Customer Relationship Management	Responsiveness	5	0.683	0.906	0.922	0.799
	Satisfaction	4	0.705			
	Assurance	5	0.698			

Divergent validity for all the construct was also examined by ensuring that the squared root of AVE of each construct is greater than their inter-construct correlations (Chin et al., 1997). Thus, it is evident that each variable is more closely and strongly related to its own construct than being correlated with other constructs (Table 2.5).

2.4.3 HYPOTHESIS TESTING

For testing the hypotheses, the final CFA model was put into SEM, and paths were drawn based upon the hypotheses framed for the study. Two models were tested in this SEM. For model one, paths were drawn for four hypotheses, namely, H1, H2, H3, and H4, which were measuring the impact of cost, convenience, status, and privacy on the adoption of IoT, respectively.

Then for model two, the impact of the adoption of IoT was measured on CRM (H5). It was observed that in model one, only cost and privacy were found to be significantly positively impacting the adoption of IoT with a beta value of 0.574 and 0.181, respectively, thus referring to the acceptance of H1 and H2 while the rest of the variables from model one are not impacting significantly. Cost, convenience, status, and privacy, in model one, together explained the variance of 35.20% (R^2_1 = 0.352). Model two refers to H5, where the impact of the adoption of IoT was measured on CRM. The standardized beta value for this relationship was found to be at 0.481 significant at a 1% level of significance. The explanatory power of this model was found to be 23.20% (R^2_2 = 0.232), which infers that due to a 100% change in the adoption of IoT, a change of 23.20% takes place in CRM.

TABLE 2.5 Correlations, Divergent Validity, and Descriptive Statistics

	1	2	3	4	5	6
Cost	**0.693**					
IoT adoption	0.301	**0.799**				
Convenience	0.433	0.194	**0.735**			
Status	0.228	0.169	0.213	**0.719**		
Privacy	0.209	0.351	0.118	0.133	**0.709**	
Customer Relationship Management	0.235	0.586	0.366	0.261	0.038	**0.894**
Mean	3.521	3.284	3.559	3.841	2.981	3.125
Standard Deviation	0.865	1.115	0.979	0.893	0.872	0.842
Skewness	−1.564	−1.208	1.071	0.869	1.218	0.875
Kurtosis	0.898	−0.196	0.450	0.448	0.165	0.586

Diagonal values with bold are squared root of AVE.

2.5 DISCUSSION

In this study, authors have tried contributing to the existing literature of CRM in the banking sector of India through testing the impact of the adoption of IoT, which is still an upcoming phenomenon in the Indian scenario. As of now, to the best of the knowledge of the authors, no study has been conducted in Indian settings which have attempted to measure CRM in the banking sector through the lens of IoT adoption. In order to measure CRM, researchers have first checked the impact of cost, convenience, status, and privacy on IoT adoption since IoT is a technological phenomenon which is completely based upon networks and the Internet thus brings in many concerns related to its cost of usage, the security of its users, the privacy of the information shared by IoT users, etc., henceforth any user would be concerned about these issues prior adoption of IoT for using banking services. Considering the aforementioned issues, the authors first tried checking the impact of cost, convenience, status, and privacy on IoT adoption before checking the impact of IoT adoption on CRM in the Indian banks.

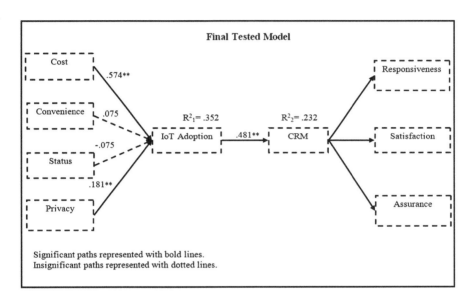

Significant paths represented with bold lines.
Insignificant paths represented with dotted lines.

In the first model, cost and privacy showed up with a significant influence on IoT adoption. Cost is found to be strongly related to IoT adoption with the beta coefficient of 0.574, which refers to the notion that customers of Indian banks are very much concerned about the cost of using IoT services prior to opting for it. The lesser is the cost; the greater is the adoption of IoT services. Rayport and Sviokla (1994) also determined two factors affecting consumer behavior, naming price/non-price factors. In addition, writers like Gupta (1988) and Mazursky et al. (1987) considered price factor as a major element in brand switching. While on the other hand, Privacy has also been found to be influencing IoT adoption with the beta value of 0.181, which shows a weaker relationship which enumerates that customers of Indian banks are not as much worried about the privacy of their data or information shared through the usage of IoT services as they are about the cost of using IoT services. Dawid et al. (2017) also mentioned that there must be some arrangement to educate the customers regarding the benefits and harms of IoT in order to increase the awareness of the customers about their data privacy. But still, when compared to other variables – convenience and status – they are little more concerned about the privacy issue of their personal information; henceforth, it can be confirmed that the greater is the privacy of users' information higher are the chances for the adoption of IoT services in the banking sector.

In the second model of hypothesis testing, IoT adoption was found to be significantly impacting CRM with the beta coefficient of 0.481, which means that through the adoption of IoT services, the customer-bank relationship is strengthening up. Ou and Sia (2003); Parihar (2012) also supported that CRM can be better performed through its factors responsiveness, the satisfaction of the customers, and assurance of the employees with the help of IoT.

2.6 LIMITATION AND SCOPE OF FUTURE RESEARCH

The present study is confined to three public and three private banks only. This study is a contribution to the extant literature regarding the adoption of IoT. The size for the sample of the study is limited to 366 only, while a large sample required money and time. Furthermore, this study is based on a total of seven variables only (four to check influencing factors and three to check the impact of IoT on CRM dimensions). The impact of IoT on CRM can be checked in other banks with other variables too. Greater sample size in other geographical areas can also be covered along with the impact of other influencing factors for the adoption of IoT can be taken for future studies.

2.7 ORIGINALITY/VALUES

Through this paper, there is an attempt to examine the adoption of the Internet of the things as well as the impact of the adoption of IoT on CRM in banks, particularly in India because being a developing country, there is a huge scope for the application of IoT in different service areas and such kind of modern technology can be a great help for it. Although, the results of this examination can also be summed up in other developing countries like Bhutan, Bangladesh, Pakistan, Indonesia, and so on.

KEYWORDS

- confirmatory factor analysis
- customer relationship management
- information technology
- internet of things
- technology acceptance model
- worldwide web

REFERENCES

Al-Fahim, N. H., (2013). An exploratory study of factors affecting the internet banking adoption: A qualitative study among postgraduate students. *Global Journal of Management and Business Research.*

Al-Hajri, S., (2008). The adoption of e-banking: The case of Omani banks. *International Review of Business Research Papers, 4*(5), 120–128.

Al-Majali, M., & Mat, N. K. N., (2011). Modeling the antecedents of internet banking service adoption (IBSA) in Jordan: A structural equation modeling (SEM) approach. *Journal of Internet Banking and Commerce, 16*(1), 1–15.

Al-Momani, A. M., Mahmoud, M. A., & Ahmad, S. M., (2016). Modeling the adoption of Internet of things services: A conceptual framework. *International Journal of Applied Research, 2*(5), 361–367.

Anwar, I., & Saleem, I., (2018). Effect of entrepreneurial education on entrepreneurial intention of Indian students. *International Journal of Research, 5*(12), 2306–2316.

Anwar, I., & Saleem, I., (2018). Exploring entrepreneurial characteristics among university students: An evidence from India. *Asia Pacific Journal of Innovation and Entrepreneurship.* https://doi.org/10.1108/APJIE-07-2018-0044.

Anwar, I., & Saleem, I., (2019). Entrepreneurial intention among female university students: A step towards economic inclusion through venture creation. In: Mrinal, S. R., Bhattacharya, B., & Bhattacharya, S., (eds.), *Strategies and Dimensions for Women Empowerment* (pp. 331–342). Central West Publishing, Australia.

Awasthi, S., (2012). Analyzing the effectiveness of customer relationship management in increasing the sales volume with special reference on organized retail stores in Indore City. *International Journal of Retailing and Rural Business Perspectives, 1*(1), 102.

Bagozzi, R. P., & Yi, Y., (1988). On the evaluation of structural equation models. *Journal of the Academy of Marketing Science, 16*(1), 74–94.

Bajaj, P., (2014). *A Study of Customer Relationship Management (CRM) in Life Insurance Corporation and Bajaj Allianz Private Limited.* Unpublished doctoral thesis, Aligarh Muslim University.

Bajaj, P., Almugari, F., Tabash, M. I., Alsyani, M., & Saleem, I., (2019). Factors influencing consumer's adoption of Internet of things: An empirical study from Indian context. *Int. J. Business Innovation and Research, 10.*

Bompolis, C. G., & Boutsouki, C., (2014). *Customer Relationship Management in the Era of Social Web and Social Customer: An Investigation of Customer Engagement in the Greek Retail Banking Sector.* Retrieved from: https://doi.org/10.1016/j.sbspro.2014.07.018.

Chang, T. Z., & Wildt, A. R., (1994). Price, product information, and purchase intention: An empirical study. *Journal of the Academy of Marketing Science, 22*(1), 16–27.

Chin, W. W., Gopal, A., & Salisbury, W. D., (1997). Advancing the theory of adaptive structuration: The development of a scale to measure faithfulness of appropriation. *Information Systems Research, 8*(4), 342–367.

Cho, Y. C., & Sagynov, E., (2015). Exploring factors that affect usefulness, ease of use, trust, and purchase intention in the online environment. *International Journal of Management and Information Systems, 19*(1), 16.

Chong, A. Y. L., Chan, F. T., & Ooi, K. B., (2012). Predicting consumer decisions to adopt mobile commerce: Cross country empirical examination between China and Malaysia. *Decision Support Systems, 53*(1), 34–43.

Cisco, (2015). *Internet of Things.* Available at: www.cisco.com/c/en/us/solutions/internet-of-things/overview.html (accessed on 21 October 2020).

Cohen, J., & Cohen, P., (1983). *Applied Multiple Regression/Correlation Analysis for the Behavioral Sciences* (2nd edn.). Hillsdale: Erlbaum.

Davis, F. D., (1989). Perceived usefulness, perceived ease of use, and user acceptance of information technology. *MIS Quarterly, 13*(3), 319–340.

Dawid, H., Decker, R., Hermann, T., Jahnke, H., Klat, W., König, R., & Stummer, C., (2017). Management science in the era of smart consumer products: Challenges and research perspectives. *Central European Journal of Operations Research, 25*(1), 203–230.

Decker, R., & Stummer, C., (2017). Marketing management for consumer products in the era of the Internet of things. *Scientific Research Publishing Inc., 7*(3), 47–70. Available at: http://www.scirp.org/journal/doi.aspx?DOI=10.4236/ait.2017.73004 (accessed on 21 October 2020).

Elkhodr, M., Shahrestani, S., & Cheung, H., (2012). A review of mobile location privacy in the Internet of things. *International Conference on ICT and Knowledge Engineering,* 266–272.

Farooq, M. U., & Waseem, M., (2015). A review on Internet of things (IoT). *International Journal of Computer Applications (0975 8887), 113*(1), 1–7.

Fayolle, A., & Liñán, F., (2014). The future of research on entrepreneurial intentions. *Journal of Business Research, 67*(5), 663–666.

Gao, L., & Bai, X., (2014). A unified perspective on the factors influencing consumer acceptance of Internet of things technology. *Asia Pacific Journal of Marketing and Logistics, 26*(2), 211–231.

Hair, Jr. J. F., Black, W. C., Babin, B. J., & Anderson, R. E., (2010). *Multivariate Data Analysis: A Global Perspective* (7th edn.). Pearson Education, Upper Saddle River.

Henseler, J., Ringle, C. M., & Sinkovics, R. R., (2009). The use of partial least squares path modeling in international marketing. *Advances in International Marketing, 20*(1), 277–319.

Hsu, C. L., & Lin, J. C. C., (2016). An empirical examination of consumer adoption of Internet of things services: Network externalities and concern for information privacy perspectives. *Computers in Human Behavior, 62,* 516–527. Available at: http://dx.doi.org/10.1016/j.chb.2016.04.023 (accessed on 21 October 2020).

Hu, L. T., & Bentler, P. M., (1999). Cutoff criteria for fit indexes in covariance structure analysis: Conventional criteria versus new alternatives. *Structural Equation Modeling: A Multidisciplinary Journal, 6*(1), 1–55.

James, E., (2009). Customer relationship management: The winning strategy in a challenging economy table of contents. *Microsoft Dynamics CRM,* 33–36.

Jerald, V., Rabara, A., & Premila, D., (2015). Internet of things (IoT) based smart environment integrating various business applications. *International Journal of Computer Applications, 128*(8), 32–37. https://doi.org/10.5120/ijca2015906622.

Jing, Q., et al., (2014). Security of the Internet of things: Perspectives and challenges. *Wireless Networks, 20*(8), 2481–2501.

Jones, R., Suoranta, M., & Rowley, J., (2013). Strategic network marketing in technology SMEs. *Journal of Marketing Management, 29*(5/6), 671–697. doi: 10.1080/0267257X.2013.797920.

Kim, J., & Lennon, S. J., (2013). Effects of reputation and website quality on online consumers' emotion, perceived risk, and purchase intention: Based on the stimulus-organism-response model. *Journal of Research in Interactive Marketing, 7*(1), 33–56.

Kim, K. J., & Shin, D. H., (2015). An acceptance model for smartwatches: Implications for the adoption of future wearable technology. *Internet Research, 25*(4), 527–541.

Kim, K. J., & Shin, H., (2015). An acceptance model for smartwatches. *Internet Research, 25*(4), 527–541. Available at: http://www.emeraldinsight.com/doi/10.1108/IntR-05-2014-0126 (accessed on 21 October 2020).

Kim, K., et al., (2014). The effects of co-brand marketing mix strategies on customer satisfaction, trust, and loyalty for medium and small traders and manufacturers. *E+M Ekonomie a Management, 1,* 140.

Kline, R. B., (1998). *Principles and Practice of Structural Equation Modeling.* New York: Guilford Press.

Kotarba, M., (2016). New factors inducing changes in the retail banking customer relationship management (CRM) and their exploration by the fintech industry. *Foundations of Management, 8*(1), 69–78. https://doi.org/10.1515/fman-2016-0006.

Krueger, Jr. N. F., Reilly, M. D., & Carsrud, A. L., (2000). Competing models of entrepreneurial intentions. *Journal of Business Venturing, 15*(5/6), 411–432.

Kuo, Y. F., & Yen, S. N., (2009). Towards an understanding of the behavioral intention to use 3G mobile value-added services. *Computers in Human Behavior, 25*(1), 103–110.

Li, X. J., & Wang, D., (2013). Architecture and existing applications for Internet of things. *Applied Mechanics and Materials, 347–350,* 3317–3321. Available at: http://www.scientific.net/AMM.347-350.3317 (accessed on 21 October 2020).

Liñán, F., & Chen, Y. W., (2009). Development and cross-cultural application of a specific instrument to measure entrepreneurial intentions. *Entrepreneurship Theory and Practice, 33*(3), 593–617.

MacCallum, R. C., Browne, M. W., & Sugawara, H. M., (1996). Power analysis and determination of sample size for covariance structure modeling. *Psychological Methods, 1*(2), 130.

Maditinos, D., Chatzoudes, D., & Sarigiannidis, L. (2013) 'An examination of the critical factorsaffecting consumer acceptance of online banking', *Journal of Systems and Information Technology, 15*(1), 97–116, DOI:10.1108/13287261311322602

Medaglia, C. M., & Serbanati, A., (2010). An overview of privacy and security issues in the Internet of things. In: Giusto, D., Iera, A., Morabito, G., & Atozori, L., (eds.), *The Internet of Things* (pp. 389–395). Springer, New York, NY.

Nasri, W., (2011). Factors influencing the adoption of internet banking in Tunisia. *International Journal of Business and Management, 6*(8), 143–160. Available at: http://www.ccsenet.org/journal/index.php/ijbm/article/view/13568 (accessed on 21 October 2020).

Ou, C. X., & Sia, C. L., (2003). *7th Pacific Asia Conference on Information Systems*. Adelaide, South Australia.

Parasuraman, A., Zeithaml, V. A., & Berry, L. L., (1988). SERVQUAL: A multiple item scale for measuring consumer perceptions of service quality. *Journal of Retailing, 64*(Spring), 12–41.

Parihar, M., (2012). Social Media: The Final Frontier in Customer Experience Management. Retrieved from: https://papers.ssrn.com/sol3/papers.cfm?abstract_id=1982325 (accessed on 21 October 2020).

Pikkarainen, K., Karjaluoto, H., & Pahnila, S., (2004). Consumer acceptance of online banking: An extension of the technology acceptance model. *Internet Research, 14*(3), 224–235.

Podsakoff, P. M., & Organ, D. W., (1986). Self-reports in organizational research: Problems and prospects. *Journal of Management, 12*(4), 531–544.

Shevlin, M., & Miles, J. N., (1998). Effects of sample size, model specification and factor loadings on the GFI in confirmatory factor analysis. *Personality and Individual Differences, 25*(1), 85–90.

Sicari, S., et al., (2015). Security, privacy, and trust in Internet of things: The road ahead. *Computer Networks, 76*, 146–164. Available at: http://dx.doi.org/10.1016/j.comnet.2014.11.008.

Suki, N. M., (2010). An empirical study of factors affecting the internet banking adoption among Malaysian consumers. *Journal of Internet Banking and Commerce, 15*(2), 1–11.

Talari, S., et al., (2017). A review of smart cities based on the Internet of things concept. *Energies, 10*(4), 1–23.

Tatnall, A., & Davey, B., (2017). The Internet of things and convenience. *Internet Research*, 353–364. http://doi.org/10.4018/978-1-5225-1832-7.ch016.

Thibaut, K., (2014). Chapter 1: Introduction. *Internet of Things-From Research and Innovation to Market Deployment*, 355.

TRUSTe, (2014). Internet of things industry brings data, but growth could be impacted by consumer privacy concerns. *TRUSTe Research*. Available at: https://www.trustarc.com (accessed on 21 October 2020).

Tsai, C. W., Lai, C. F., & Vasilakos, A. V., (2014). Future internet of things: Open issues and challenges. *Wireless Networks, 20*(8), 2201–2217. Available at: http://link.springer.com/10.1007/s11276-014-0731-0 (accessed on 21 October 2020).

Venkatesh, V., Morris, M. G., Davis, G. B., & Davis, F. D., (2003). User acceptance of information technology: Toward a unified view. *MIS Quarterly*, 425–478.

Wheaton, B., Muthen, B., Alwin, D. F., & Summers, G. F., (1977). Assessing reliability and stability in panel models. *Sociological Methodology, 8*(1), 84–136.

Worthington, R. L., & Whittaker, T. A., (2006). Scale development research: A content analysis and recommendations for best practices. *The Counseling Psychologist, 34*(6), 806–838.

Yu, X., Nguyen, B., & Chen, Y., (2016). Internet of things capability and alliance: Entrepreneurship orientation, market orientation, and product and process innovation. *Internet Research, 26*(2), 402–434. doi: 10.1108/IntR-10-2014-0265.

CHAPTER 3

Cross-Cultural Management: Opportunities and Challenges

KAVITA THAPLIYAL[1] and MAHENDRA JOSHI[2]

[1]*Associate Professor and Chief Corporate Trainer (Business Communication and Soft Skills), Amity International Business School, Amity University, Noida, Uttar Pradesh, India*

[2]*Associate Professor, Seidman College of Business, Grand Valley State University, Michigan, USA*

ABSTRACT

The management of people, material, and culture from a diversified background is called cross-culture management. Different cultures may originate in a family, neighborhood, state, country, and organization. We all tend to think, behave, and live with certain values, beliefs, norms, attitudes, eating, speaking, and dressing patterns, which we then follow normally as a routine in life. Cross-culture refers to people who differ in origin, religion, geography, nationality, race, ethnicity, age, language, gender, and sexual orientation. People of different cultures, when meet, their way of speaking, behaving, working, eating, and living all differ, and this may lead to disparity in opinion and understanding between them. If both individuals and parties learn to accommodate, adjust, respect, and listen patiently to each other's viewpoints and behaviors, then there are fewer problems, and everything becomes amicable between them. When one assumes his culture is superior to others, then clashes or differences occur. This occurs mainly when one has minimal knowledge and cultural understanding of the world or sheer arrogance and attitude on one's part. Cross-culture management plays a vital role here in the efforts to bridge the cultural gaps between two or more people, parties, organizations, and countries by trying to negotiate, mediate by gestures, goodwill, etiquette, behaviors, and languages. Communication plays a very important and strategic part here and is the fundamental link between the two entities. Cross-culture communication originates when people from different social, cultural, and geographical backgrounds try to converse with each other in similar or different ways. So as we learn to communicate, we also learn about each other's culture in close tandem, influencing our behavior and the words we speak.

3.1 CROSS CULTURE BUSINESS ENVIRONMENTS

Today rapid globalization and connectivity have shrunk the world. Advance logistics, fast internet connectivity, and shorten travel time has helped in making the world a small place. This interconnectivity with the outer world is exciting and encouraging. It opens a plethora of opportunities in our personal and professional life. Businesses and organizations have also grown in tandem with technological and geographical advancement. With all the advancement and changes, it is perceived that cultural barriers have vanished, and the world is one close family. This is far from being true; the truth remains that still, the world remains a divided force with ample differences in culture, values, views, and communication.

In cross-culture management, these gaps play a vital role, and one has to carefully master them to succeed and sustain in overcoming the prevailing difficult and diverse global barriers. One must be open to accept and learn the nuances of diverse cultures and their business practices and norms. In order to gain their acceptability and survive in their markets, organizations have to adapt quickly to the global changes and work on strong cross-cultural teams in tandem. This is challenging and taxing even for the best firms, as their managers having ample international exposure often find immense difficulty to adjust to the change in working with multicultural teams and business environments. This happens due to the non-adaptation of global work practices in your working environments, which can lead to conflicts, misunderstandings, stress, and loss of business opportunities. Companies are fast changing their outdated attitude in order to survive global markets and scale their reach. It's becoming a prerequisite for international businesses to have cross-cultural competence and quick adaptations to local business practices. With a high reputation and finances at stake in world markets, managers are learning to please and gel with their colleagues and customers, thus reacting quickly to their changes in interests and tastes. Even in adverse working conditions, managers are overcoming linguistic barriers, the difference in opinions, misunderstandings to compete globally by forging multicultural teams. Today employees speak multi-languages, learn diversified cultures by taking a keen interest in varied dressing and eating styles. This is to make the business environments more soothing and approachable, thus ensuring there is no loss of business with international clients.

3.2 MANAGING GLOBAL TEAMS

Cross-culture, management becomes all more important as its effective implementation helps individuals, teams, and organizations in attaining a deep overview of multicultural market environments, global outreach, and sustainability. Seeing today's disruptive economy and volatile markets, one has to tread their business paths carefully with cultural sensitivity and avoiding ethnocentrism, parochialism, cultural shock to operate smoothly from any area of the world. With the global business landscape growing at a faster pace, multi-skilled leaders are needed to evolve the organizations and garner them with international trade practices and global compliance norms. This will help them in edging out the competition, imbibing cross-cultural relationships, and constructing winning strategies by efficiently integrating international business perspectives. Managing cross-cultural teams provides international exposure and innovative thinking to work across time zones and enhance global competitive edges. International companies have to practice cultural ethics with multi-dimensional value systems.

Every company has its own value systems, communication, and management style known as its corporate culture. It can be a set of beliefs, patterns, values, symbols, ceremonies, and styles of operating business, followed by management and its employees. It can also be defined as the development of moral, intellectual, and behavioral faculties within the working environment based on the attitudes, priorities, and beliefs of the workers. It lays the foundation of how things are practiced in the organization and not just stated, though there is an intact system of policies and strategies that channelize the workflow. Corporate culture is subtle, intangible, and very hard to quantify, but it is the right mantra to condition the actual behavior of people. The management style of the organization is influenced and shaped by numerous factors like leadership, organization, motivation, and communication approach and methods. However, culture defines the working patterns of the organization. Each organization has its own distinct culture or set of values. Most organizations do not intentionally try to create a certain culture, and some consciously try to do so. The culture of the organization is characteristically created unconsciously, based on the morals and values of the top management or the founders of an organization. It signifies their style of doing business and principles for adherence at work.

Companies are extremely conscious of their work culture and work hard by spending lots of energy to maintain that, often training their employees and managers for the same. It the fact that many successful companies are known for their culture, and their growth and success reflect in accordance. Corporate culture can be based on hard work, a sense of community, and even respect for others. It can reflect on the company's hiring patterns, where they are often looking for particular traits, often ignoring hard skills, backgrounds, and job categories. Some look for a positive attitude, humor, team player, or people who can lend themselves to others or causes. Their hiring reflects the team or workforce they visualize for their organizational culture. Spending ample time with your team and communicating effectively with your people also reflects corporate culture. This makes the workforce feel important, happy, motivated, and inclined to company roots, further reflecting in their professional and organizational growth. Vice-versa, it radiates that the company is proud of its employees as individuals apart from them being their employees. Stakeholders and top management are vigilant and maintain their corporate culture safeguarding their ethical values with clearly defined rules and boundaries. However, it is not mandatory and differs pragmatically from company to company with a diversity of cultures and boundaries.

3.3 LEARNING OUTCOMES

This chapter on Cross-cultural management will give you a comprehensive multi-dimensional approach to lead teams across geographical boundaries. It will reflect on the skill sets imbibed by successful companies and leaders on scaling unmatched heights of behavioral and cultural diversities and bridging the gaps through ethical human integrations for creating exceptional value for their organizations. It will also provide a detailed insight into today's global multicultural business communities, and their flexible, evolving leadership is working patterns to counter-culturally sensitive operational environments. Through this chapter, we will try to broadly amalgamate why corporate culture is a synthesis of management, values, and communication styles. It will display and introspect how global teams manage differences in opinions of colleagues and customers in lieu of their cultures, practices, and preferences in an international business context. The chapter will further enhance the manager's capacity to increase confidence, collaboration, effectiveness, and consistency among global teams and develop a deep understanding of cultural perspective differences and the knowledge to overcome and navigate the complexities of international business organizations.

3.4 CROSS-CULTURAL MANAGEMENT

3.4.1 UNDERSTANDING CULTURE AND ITS DIVERSITIES

Culture is one of the most required conventional skills for dealing in international business articulations and arrangements. Culture affects three major areas that influence people Communication, thinking, and behavior. Having a detailed understanding of culture gives an advantage in all kinds of business dealings and negotiation. When a difference in opinion among individual occurs, it gives set back in a low capacity, but when the same occurs in an international perspective, it creates an enormous barrier that obstructs or absolutely impasse the negotiation process.

For an effective business negotiator, skilled, and experienced, it is very important to have global cultural awareness and understanding. Cultural mindfulness or awareness is the principle of communication that enables us with cultural values, believes, and perceptions. It tells us how do we perceive certain things, see the world, and react in a particular way. We also notice that cultural awareness befits dominance when we have to deal with people from different cultures. It is a common phenomenon that people perceive, understand, and evaluate things in a different way. It can be understood in a way that certain beliefs or behavior may be suitable in one culture, but it might not be appropriate in another setting. Misunderstandings might also arise when we might add our own perception and meaning to others' viewpoints.

3.5 LEVELS OF CULTURE

Culture is traditionally observed through two techniques:
1. **The Psychological or Psychic Level:** In this form of culture concentrates on the assumed norms, behavior, and attitudes of individuals compared to a particular culture or differentiation between two cultures.
2. **The Institutional Level:** This form of culture is looked at from a massive level rather than an individual. In this form, we perceive culture from a society, country, nation, community, group, etc.

It has been observed that people born and grew up in the same family; the country generally shares similar cultural traits and characters. Within organizations, states, and country there may be a possibility that they might have a difference in opinion, but when it comes to nation and representing the country, they tend to be united and bonded.

One of the foremost trends of culture is that it formed with levels and sublevels. We can generally divide them as nation, religion, organization, team, and individual. The widest form in the nation and the smallest is the individual having tangible and intangible sublevels.

3.5.1 EDGAR HENRY SCHEIN MODEL OF CULTURE

According to Edgar Henry Schein, a former professor at the MIT Sloan School of Management has introduced a model of organizational culture. His model describes that there are three prominent layers of organizational culture first-artifacts, second-values, and third basic assumptions and beliefs (Figure 3.1).

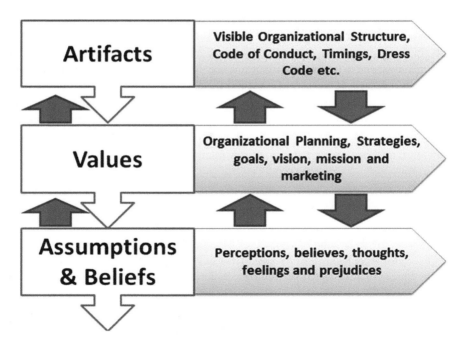

FIGURE 3.1 Edgar Henry Schein's organizational model of culture-3 levels.

Organizational culture is a nonfigurative thought and consequently challenging to appreciate. Keeping it in mind, in 1980, Professor Edgar Henry Schein developed a model known as the organizational cultural model that demonstrated the culture within an organization with much clarity and explained through three levels artifacts, values, assumptions, and beliefs that determine how organization perceive cultural change.

Artifacts are the basic codes and symbols that show the basic trade of an organization. They include-logos, brand image-dress code, architect, ambiance, layout. These symbols are not only symbolic of internal channels but also of external organizational networks.

Every organization has a value system based on certain rules, codes of conduct, and standards. This includes organizational planning at various levels, strategy, policy-making, internal, and external affairs, goals, vision, mission, and promotional policies and branding and handling organizational reputation and stake. Every organization has its important conventional assumptions, believes, perceptions, and thoughts that help the organization to formulate certain principles and value systems and is known for the same.

3.5.2 *HOFSTEDE'S CULTURAL DIMENSIONS THEORY*

The theory of Hofstede's cultural dimensions institutes an outline orbiting around cross-cultural communication, which was devised by Geert Hofstede. The dimensions collectively portray the impact of the culture ingrained in society on the values of the members of that society. They also describe the relationship between these values and behavior, with the help of a structure based on factor analysis. In other words, this theory studies significant aspects of culture and provides them a rating on a comparison scale.

Geert Hofstede developed a unique framework that describes the relationship between cross-cultural communications. The dimensions describe the significance of culture depending on the values, behavior. The relationship between values and behavior are measured based on factor analysis, which describes as Hofstede's cultural dimension theory. This theory is a systematic study based on the worldwide survey of IBM employees. Originally this theory was conceptualized with four dimensions, but with time has framed six dimensions such as: power distance index (PDI), individualism vs. collectivism (IDV), uncertainty avoidance (UAI), masculinity vs. femininity (MAS), long-term orientation vs. short-term orientation (LTO), and indulgence vs. restraint (IND) (Table 3.1):

1. **Power Distance Index (PDI):** It is described in two forms; first high power distance and low power index. In high power, index show inequality and visibility of power differences, rand, position, stature, is given preference and reassures bureaucracy. On the other side, low power distance encourages participative involvement at the organizational level.
2. **Individualism vs. Collectivism (IDV):** This is one of the dimensions in which societies are taken into consideration on two different attributes. First individualism-were people are reserved to themselves and the factor 'I' is predominant. Here personal preferences are taken first into consideration. On the other side, Collectivism is collective behavior psychology, where people love to work in a team, and together they work, and preferences are towards the society and well-being of everyone. The word 'WE' is more than the work 'I.'
3. **Uncertainty Avoidance (UAI):** This index explains the level of tolerating uncertainty and ambiguity. They are generally in two categories-High uncertainty and low uncertainty. In high uncertainty describes low tolerance towards uncertainty and risk-taking. It is dealt with strict rules and compliance. On the other side, low uncertainty describes a high tolerance for risk-taking. Mysterious things are accepted with less formality and regulations.
4. **Masculinity vs. Femininity (MAS):** This indicator, as clearly shows, describes the society role towards males and females. It is also indicated as 'hard and soft. In masculinity society, strong egos, attitude, pride, and authoritativeness are embodied, whereas, in feminist's relationship, responsibility, family more is taken into consideration.
5. **Long-Term Orientation vs. Short-Term Orientation (LTO):** Society has a time framework where long orientation indicates long-term goals, patience, and persistence towards goal orientation. On the other side, short-term orientation is a quick response and focuses on day-to-day needs and association.
6. **Indulgence vs. Restraint (IND):** Indulgence means the absorption or fascination of society towards desires and impulses. Every society has an impulse to life happily, enjoy, and have a safe environment; on the other side, restrain is the restrictions and control checks that suppress the needs of individuals and restrict them towards norms and regulations.

TABLE 3.1 Hofstede's Cultural Dimension Model

Cultural dimension	Definition	Examples
Power distance	**Power distance** is the extent to which the less powerful members of institutions and organizations within a country expect and accept that power is distributed unequally.	**Low**: U.S. and Canada **High**: Japan and Singapore
Individualism and collectivism	**Individualism** describes cultures in which the ties between individuals are loose. **Collectivism** describes cultures in which people are integrated into strong, cohesive groups that protect individuals in exchange for unquestioning loyalty.	**Individualistic**: U.S., Australia, and Great Britain **Collectivistic**: Singapore, Hong Kong, and Mexico
Masculinity-femininity	**Masculinity** pertains to cultures in which social gender roles are clearly distinct. **Femininity** describes cultures in which social gender roles overlap.	**Masculinity**: Japan, Austria, and Italy **Femininity**: Sweden, Norway, and the Netherlands
Uncertainty avoidance (UAI)	**Uncertainty avoidance** is the extent to which the members of a culture feel threatened by uncertain or unknown situations.	**Low**: Singapore, Jamaica, and Denmark **High**: Greece, Portugal, and Japan
Confucian dynamism	**Confucian dynamism** denotes the time orientation of culture, defined as a continuum with long-term and short-term orientations as its two poles.	**Long-term**: China and Japan **Short-term**: U.S. and Canada

Source: Hofstede (1991, p. 28).

In today's scenario, international business is concerned, as one of the dimensions of culture as an important facet. With the help of Hofstede's Cultural Dimension Model 1 can easily understand the societal structure, difference in culture, power, influencing attributes, policy, norms, and regulations prevailing in the society and how the knowledge of them will help the budding learners to successfully nautical the globe and its international markets.

Today we face innumerable challenges at both national and global levels, such as communication, understanding, negotiation, and relationship management. Through Hofstede's six-dimensional models, we can help ourselves relate those challenges and further build our human relations practices the same.

3.5.3 THE LEWIS MODEL OF CROSS-CULTURAL UNDERSTANDING

Renowned cross-cultural expert Richard Lewis designed a unique model of cross-culture and an author of a famous book, "When Cultures Collide: Leading across Cultures" published by the Wall Street Journal in 1996 is also known as "an authoritative roadmap to navigating the world economy."

According to the Lewis Model, countries can be categories into three wide categories:

1. **Linear Active:** The people belonging to this culture have a unique pattern and lifestyle. They are punctual, time-bound, systematic, organized; believe in doing one thing at a time. Countries like: Germany, the USA, and Switzerland are the best examples of this category.
2. **Multi-Actives:** People belonging to this culture are talkative, influencing, and multi-tasking. They plan a lot but do not follow the planning schedule and are flexible. Countries like: Latin America, Italy, Arab is the best illustrations.
3. **Reactive:** In this category, people never react until they are provoked or criticized. Their nature is generally loving and caring. They are good listeners, have respect and pride. Their value system is rooted and is hard working. Countries in this category are China, Japan, Korea, India, and Vietnam (Figure 3.2).

3.6 CROSS-CULTURE

Cross-culture refers to interaction, dealing, and comparative study or analysis between two or more cultures. When applied with a broader Management perspective in a detailed cross-cultural context, it is referred to as cross-cultural management, which imbibes organizational behavior (OB) and International Human Resource as an integral part of the study. Cross-cultural Psychology, Anthropology, and Sociology are interdisciplinary contributing verticals. Organizations in business terms define it as an influencing influx of societal ethics, practices, and cultures in managing day-to-day global working teams across cultures and nations. When measured in an individual context, it outlines one's core values, personal reactions, experiences, and cognitive structures prominently.

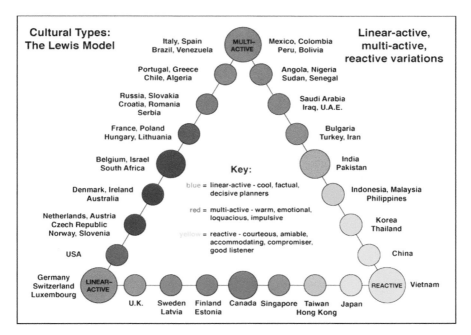

FIGURE 3.2 The Lewis model of cross-culture.
Source: https://www.crossculture.com/about-us/.

3.7 CROSS-CULTURAL MANAGEMENT

Global organizations have to adapt and modify as per complex, volatile, disruptive, and diverse ever-changing markets to compete and excel in a level playing field. The major focus area of cross-cultural management is to effectively and ethically adjust to the difference in opinion, situation, and decisions arising due to varied customs, cultures, standards, laws, and practices in the socio-economic system. With the economies growing at an ever-faster pace matching the liberalization and technological advances, organizations are crossing boundaries for gaining global footprints. While companies find their new overseas explorations and ventures lucrative, it is actually far more complex due to the influx of numerous diversities in cross-culture amalgamations. These challenges, imbalances, and differences have formulated the study of cross-culture understanding and adaptation, which supports global organizations with relevant strategic options for managing day-to-day complex working hindrances. We can define cross-culture management as a set of varied assumptions, presumptions, expectations, introspections, negotiations, formulations, and adaptations. In addition, it is a fact that we cannot isolate business management from cultural management due to differences and similarities in societies. Somewhere a work-life balance has to be made in our professional and personal lives without misbalancing either. The deep imbibing of human values affects human attitudes, human ideology, and even virtual and technological transfers. This cross-cultural diversity also influences the business processes, employee behavior, decision making, and communication styles. Therefore, it is imperative for organizations and managers to clearly understand and accept the vital importance of cross-cultural integrations.

3.8 IMPORTANCE OF CROSS-CULTURAL MANAGEMENT

Fast development, migration, and globalization are affecting business environments rapidly, and multi-functional managers must adapt quickly to such working patterns. Growing competition, innovative technologies, and integrated networks are defining globalization. This multicultural business environment diversity is also bringing in working and functioning complexities. Economic advances, financial market fluctuations, and technological changes are affecting different aspects of culture. Global managers need compatible and effective solutions to tackle these hindrances, along with legal, political, and historical characteristics. Here the vital role of cross-cultural management comes into prominence in taking ethical and preventive actions to solve any conflict in the best amicable way. Multicultural organizations need dynamic managers to overcome complex situations and implement an array of decisions every passing day. These plethoras of decisions in such highly complex international environments include managing multiple offices across multiple global locations, recruiting multicultural teams, working in different time zones, deep understanding of multi-culture environments, handling resources and logistics from remote locations, initiating joint ventures, mergers, and acquisitions, overseeing compliances and due diligence, work with virtual teams, handling conflict resolution, administrating corporate culture and ethics, corporate communication, appraisals, and training and development. For this, managers must be adequately trained with detailed and effective knowledge management, global HR practices, international law, broader outlook, and pragmatic bicultural skills. To enhance the competitive position of global firms and scale growth, global teams must be nurtured with innovative thinking, out of the box ideas, and encouraging thought processes. Cultural misunderstandings, egos, and differences in opinion can severely hamper the projects undertaken, interfere, and create barriers in day-to-day working environments. To achieve successful completion of ventures, leadership has to be flexible, patient, culturally inclined, and extremely sensitive to effectively counter-cultural differences and social misunderstandings. Organizations have to introspect deeply by working on organizational behavior and human value systems that include human feeling, thinking, and acting. This helps in forecasting positive corporate health and happiness index, which is vital for organizational survival and sustainability.

3.9　KEY CROSS-CULTURAL BARRIERS AND SOLUTIONS

The major cross-cultural roadblocks are languages, technology, beliefs, stereotypes, interpretation, trust, perception, parochialism, ethnocentrism, bias, ego, borders, values, customs, patterns, traditions, and social norms. Some appropriate solutions to these barriers are learning the local language, land, culture, eating, and dressing habits. Hosting social and cultural events, interaction sessions, sports events, outings, and family meets are some of the effective and easy ways of bridging the multicultural gaps. Multinational organizations hold culturally oriented seminars, conferences, and theme parties for staff. Trainings in soft skills, technology, forex, international trade, image management, etiquette, and corporate social responsibility (CSR), and cross-cultural communication are imparted to staff holding key positions in sensitive business environments. Managers who are also key negotiators must tread carefully while communicating and closing deals in cross-cultural business environments. They must not rush the deals in order to be efficient or save time, thus hampering the negotiation process and eventually losing the deal. Cultural views, client behavior, business etiquettes, and communication styles make a forceful impact while discussing key professional parameters and also while handling conflict resolution.

They are to be strictly kept in mind and adhered to while conducting fruitful cross-cultural negotiations and concluding business agreements to avoid any potential misunderstandings. During such intercultural communication, it is vital to study the person involved and the context of his communication, as it is often hard to understand the thoughts and behavior he wishes to express. So intercultural competence is important during cross-cultural communication to outline the inherent and unseen values behind the noticeable manners he is expressing while negotiating.

3.9.1 PAROCHIALISM

It is the state of being narrow-minded, selfish, or bias towards one's own self and culture. In business terms, it emphasizes harping on the local perspective of working and living in confined, limited demarked boundaries. You can define it as the process of viewing and judging the world solely by your own eyes, leaving limited scope for outward interventions. Such people are self-centered, having restricted and parochial attitude towards the thought processes and culture of others. They are not willing to accept the diversity of living and working patterns of other people and their teams. It is one of the most major obstacles that managers face while handling global working teams. Global managers should counter the same by showing a broader outlook and value multicultural team's suggestions and views positively. They should shed rigidness and attitude by involving them fully and transparently in achieving organizational objectives and goals. It is vital for them to fully understand and adapt to the foreign culture and the law of the land to tide over these complex hindrances.

3.9.2 ETHNOCENTRISM

This thought promotes the division of culture, races, religions, and ethnicities by believing in the superiority of one culture over the other. Individuals seized by ethnocentric behavior think they are superior to others due to their heritage, race, and rich bloodline. Actually, they have a culturally ignorant mindset due to limited exposure to diverse cultures, which has made them believe that their own way of life is natural or correct. While many cultures may be identical or overlapping in concepts and formations, many cultures may differ completely in their outlook. Some extremists are in a shock to absorb or acclimatize these differences and behave differently, displaying varied reactions of bias behavior. At times, such individuals or groups might try to forcefully influence other individuals or groups by converting their ways of living to their own accordance. This fanatical aggressive behavior can be repulsive and lead to major conflicts, genocide, and war. Managers must control their instincts and reactions towards varied human behavior by being culturally relative and learn to ignore trivial issues in workplaces. A fair division of key positions in the organization as per location and market demand is one of the apt solutions to this problem.

3.10 BUSINESS PERSPECTIVES OF CROSS-CULTURAL ORGANIZATIONS (Figure 3.3)

FIGURE 3.3 Business perspectives of cross-cultural organizations.

3.10.1 GLOBAL BUSINESS ENVIRONMENTS

Multinational enterprises (MNE) operating in foreign lands majorly defines global business environments and are the primary stakeholders in forging manufacturing, sales, and services there. The major factors influencing foreign companies in local land include its topography, state, and political governance, religion, and culture, visa, and tariffs, legalities, and taxation, and the overall economy.

3.10.2 INTERCONTINENTAL STRATEGY

Every company's commercial footprints are reflected by their core strategies for different geographical markets in accordance with the prevailing native entities. That outlines that every market and its functioning differs as per its topography, and their goals and profit are measured in lieu. OB, Corporate Culture, and International Human Resource Management (IRHM) are defined in accordance.

3.10.3 EXPATRIATE MANPOWER AND LOCAL HUMAN CAPITAL MANAGEMENT

Global staff recruitment and placement place a vital role in multicultural working environments. Relocation, visa, logistics, and settlement of staff to a foreign land are well taken care of by respective companies.

3.10.4 INTERCULTURAL MENTORSHIP

Global business leaders are evolving organizations with multifaceted core competencies across product verticals, business domains, and countries by encouraging igniting young professionals for ideation, incubations, and startups—mentoring and leading global and virtual teams across different time zones.

3.10.5 FORGING INTERNATIONAL TIE-UPS AND PARTNERSHIPS

International organizations, when scaling their international business and global competence, forge new alliances, partnerships, mergers, acquisitions, and compliances in respective countries to strengthen their local base, market playing field, strategic support, legal obligations, and create winning teams.

3.10.6 CROSS-CULTURAL COMMUNICATION

Verbal and non-verbal interactions are based on linguistic and cultural business influences—resolving common and conflicting commercial interests through amicable thought processes, actions, and mutually beneficial agreements.

3.10.7 MULTICULTURAL CONFLICT RESOLUTION

The swift solution to disagreements through cordial negotiations is the key to overcome cultural barriers; otherwise, they can snowball into major controversies severely affecting the growth and, at times, the existence of multicultural organizations. Intercultural acceptance and adaptability play a strategic role here.

3.11 MAJOR CROSS-CULTURAL FUNCTIONS IN ORGANIZATIONS

1. Defining organizational policies and work culture as per global outlook.
2. Amalgamating HR procedures amidst corporate entities in diverse nations.
3. Recruiting candidates with broader perspectives and insight on multicultural environments.
4. Training employees on soft skills diversified behavioral development and cross-cultural communication.
5. Facilitating and empowering cross-cultural teams.
6. Handling conflicting regulatory environments for business.
7. Create a transparent paradigm structure for success within the organization.

3.12 ADVANTAGES OF CROSS-CULTURAL DIVERSITY IN GLOBAL BUSINESS ORGANIZATIONS

The globalization of the workforce is considered one of the most positive aspects of world trade. Today cross-culture management complements global business operations as it nicely interprets how employees work together and understand others better by focusing on the human values and behaviors of the ecosystem. It has been observed that organizational growth increased manifold within diverse and efficient multicultural working environments. Highly managed cross-cultural teams seemed to enhance all the performances within and outside the working zones of professional parameters. Organizations with intercultural teams have competitive advantages over other firms in terms of product innovation, advertising, service orientation, communication, negotiations, and shortlisting and assessment of foreign distributors, clients, and partners. This diversity of orientation, gender, sex, age, behavior, culture, and origin within the workforce is an added advantage for any positive organization.

3.12.1   GLOBAL EDGE IN BUSINESS

Efficiently handled multicultural work teams are leading the business domains in complicated global business environments. They are better trained to communicate and engage proficiently in intercultural working scenarios. When such companies are in global expansion mode to scale operations, then again, their cross-cultural workforce is better placed to acclimatize in diverse conditions as they are well versed due to prior avid exposure in foreign domains. They are also better placed on legal and compliance parameters.

3.12.2   INTRA DEVELOPMENT OF ORGANIZATIONAL WORKFORCE

A rich pool of talent accumulates with a diverse workforce encompassing the knowledge base of multinational organizations. There are wide arrays of skills, including innovation, knowledge, talent, creativeness, local intelligence, and exposure, which exchange hands in formulating winning strategies to crack difficult deals with positivity.

3.12.3   HIGHLY MOTIVATED AND ADEQUATELY TRAINED TEAMS

The staff of global organizations is high on moral support as their company has good overseas credentials as a multinational company. They are also adequately trained by their company in soft skills, image management, foreign language, and cultural adaptation through regular management development programs. Workers' performance levels are better as they take immense pride in involving themselves in the company's work ethics, achievements, objectives, and growth initiations.

3.12.4 LOYALTY AND LIFETIME ENGAGEMENT

Emotional commitment towards organization is more in intercultural companies, and that reflects in their working engagements and patterns. Such organizations also engage their employees with care, benefits, and personal support, which keeps them deeply connected, and vice-versa encourages them to give their best for achieving organizational objectives. By engaging their employees meaningfully, companies also tend to influence customer relations efficiently by creating loyalty and reward systems for them too. Highly engaged working teams to understand the true value of the brand they represent, and they work motivated to enhance the client's overall experience. Such companies and their staff rarely want to leave each other during their professional working span and conduct many exit interviews to judge the mindset of the distressed staff member.

3.12.5 BETTER ORGANIZATIONAL STRUCTURE

We find a better system and structural diversity in diverse global working teams. Their proficient and experienced management system has better command and control over operational systems and strategic overviews. There is a clear demarcation in the span of control, a delegation of authority, and standard operating procedures.

3.13 THE ADVANTAGES OF CROSS-CULTURAL DIVERSITY IN GLOBAL BUSINESS

In the fastest global economy, the globalization of the workforce is a positive reinforcement that gives support to organizations, and the significance of cross-cultural management help in developing remarkably in global business operations by understanding diversified culture and human psychology across borders. It not only gives us an outlook to understand others' cultures but also helps in bridging the gap by elucidating how others work in building organizational mission and association. The cross-culture experts believe cultural diversity in the workforce always helps in enhancing work productivity and effectiveness, and therefore effective cross-cultural knowledge is becoming highly important in business decisions and global networking. It is also very essential to notice that cross-culture helps in discharging various responsibilities, which are discussed in subsections.

3.13.1 CROSS-CULTURE ADVANTAGES

1. **Global Expansion:** With the expansion of globalization, the global boundaries are shrinking, and businesses are expanding in increasing number and size. A few decades ago, global expansion to business was only limited to personal meetings and networking with the outer world, but today technology has given wings to understand the global culture, and one can connect and expand business sitting in one country to the nooks and corner of the world using technological tools for connection. Cross-cultural understanding helps individuals and organizations in reducing cultural stereotypes and prejudices; it further results in network better with global communities. Therefore, organizations must inculcate cross-cultural diversity for better global expansion and understanding.
2. **Development of Workforce Skills:** Due to the disruptive economy and changing global workstations, it is becoming extremely important to have sound knowledge and techniques to enter foreign markets. The methodology that one uses to adopt in the past is not sufficient to survive today. Organizations need to understand cultural upbringings can affect people's decision-making, communication, relationships, handling others, also leading, and controlling teams. Differences in demographics and economic structures are antagonizing developing countries to bridge the skill gaps. Cross-cultural diversity has many benefits to youth; they can enhance their productivity, understanding technology, innovations, and creativity by connecting with global brains through accepting cross-cultural sensitivity and learning. Cross-cultural diversity is one of the prime factors that are deeply rooted in value system beliefs and rules that are reproduced in our everyday behavior.
3. **Morale Support and Motivation:** Cultural integration is one of the most influential factors that help in increasing employee morale and confidence. Organizational cultural integrity and diversity encourage the staff to be connected with each other resulting in better performance, harmony, and productivity. It has also been noticed that

training and development programs on cross-culture brings more awareness and adaptation towards others culture and thus result from respecting towards other community and religion.

Some other features of cross-cultural advantages in global business are:

1. Product development according to global needs;
2. Service enhancement keeping customer awareness in mind;
3. Interaction and communication with international communities;
4. Identifying overseas distributors, networking, and building relations;
5. Preparing marketing strategy and advertisement campaigns;
6. Negotiations for global partnership and international business deals.

3.14 INTERNATIONAL COMPANIES ADOPTED CROSS-CULTURAL UNDERSTANDING

Cross-cultural understanding helps in amalgamating local and global needs for a successful outcome. There have been several examples where companies have mastered local + global aspects and thus are expanding in international markets having globalization in mind. Google, Starbucks, McDonald, Nestle, PepsiCo, KFC, Apple, Samsung, HUL(Hindustan Unilever Limited), Ford, Lays, Gillette, Huggies are such great illustrations that have adopted the local needs of culture and mingled with the local needs keeping global markets in mind. Amazon, Google, when reached India, initially found a tough time to establish its place; Google then went local and changed its appearance in India having a changed appeal from www.google.com to www.google.co.in for India. Many eatable brands have adapted Indian flavor to survive in India; similarly, all food courts shut down in the gulf five times a day due to cultural respect and harmony. TV Channel in India ZEE Network is also very popular globally because it integrates global culture. Indian Prime Minister Mr. Narendra Modi is very popular globally because of his cross-cultural understanding, adaptability, and flexibility of cross-cultural association.

3.15 CONCLUSION

We hereby understand that cross-cultural amalgamations are immensely significant to organizational growth in this fast-paced era of fast-changing global business orientations. It not only formulates the organization's expansion strategies but also helps to engage employees from diverse backgrounds constructively and gives them a competitive edge and sense of pride over rival firms. Cross-cultural management also handles and minimizes operational issues like uncertainty, conflicts, engagement, motivation, performance, and organizational objectives efficiently. It also counters the challenges of multicultural working environments like individualism, ethnocentrism, parochialism, and cultural distance subtly. The unconventional ways of handling cultural differences in organizations have proved futile and abysmal in today's context. Companies must evolve every day to overcome these complex barriers with innovative human capital engagement strategies keeping the present global economies in mind. They must integrate "our way with their way "concept as a gateway model to tide over the turbulent oceans of cross-cultural diversities.

KEYWORDS

- **cross-culture**
- **etiquettes**
- **individualism vs. collectivism**
- **multinational enterprises**
- **power distance index**
- **uncertainty avoidance**

REFERENCES

Emad, A. A. I., (2016). *The Cross-Cultural Management and its Relevance in Supporting the Global Business Operations*. https://www.academia.edu/26731769/The_Cross_Cultural_Management_and_its_relevance_in_supporting_the_Global_Business_Operations (accessed on 21 June 2021).

Hofstede, G., (2001). *Culture's Consequences: Comparing Values, Behaviors, Institutions, and Organizations Across Nations* (2nd edn.). Thousand Oaks, CA: Sage.

Hofstede, G., et al., (2002). What goals do business leaders pursue? A study in fifteen countries. *Journal of International Business Studies, 33*(4), 785–804.

Jolita, G., & Rasa, D., (2010). The growing need of cross-cultural management and ethics in business. *European Integration Studies, 4*, 148–152.

Kavita Thapliyal & Mahendra Joshi (n.d.). International culture. In: *The Environment of International Business* (pp. 129–158). Retrieved from: http://www.unice.fr/crookall-cours/iup_cult/_docs/_RUGM_Chapter-05.pdf (accessed on 21 October 2020).

Mohammadian, D. H., (n.d.). *An Overview of International Cross-Cultural Management*, 25.

Thomas, D. C., (2016). *Cross-Cultural Management*. Retrieved from: https://www.oxford-bibliographies.com/view/document/obo-9780199846740/obo-9780199846740-0074.xml (accessed on 21 October 2020).

CHAPTER 4

Exploring the Relationship Between Personality and Work-Life Balance

JOSHIN JOSEPH

Jr. Accountant and ICWAI Trainee, Kerala Minerals and Metals Ltd. (A Govt. of Kerala Undertaking), Kerala, India

ABSTRACT

This study explored the relationship between the psychological characteristics of the employee and work-life balance (WLB) with the help of data obtained from 520 service sector employees from the state of Kerala. The employee personality was measured with the help of the Ten Item Personal Inventory (TIPI) developed by Gosling, Rentfrow, and Swann (2003), a small-scale replica of the big five models of personal psychology. With the help of general linear modeling (GLM), it has found that employee personality characteristics have a significant as well as a large effect on the employee WLB level. The study reveals that the WLB is a phenomenon, which is interlinked with the psychological characteristics of the employee, and therefore, the interventions at a psychological level can have the potential to enhance the employee WLB level.

4.1 INTRODUCTION

Several researchers (e.g., Brough et al., 2014; Kalliath and Brough, 2008; Valcour, 2007; Greenhaus and Allen, 2006; Voydanoff, 2005) have identified the psychological interlink of the employee WLB. The word personality is derived from the Latin word 'persona,' which in turn means 'mask.' That is, personality is the mask that the individuals present before others (Sociology Guide.Com, 2016). It is dynamic and gets updated lively in accordance with the learning and experience of an individual with his living environment.

Personality can't be either purely defined biologically or psychologically (Hutton, 1945). It is because of these characteristics, personality is often termed as the study of 'nature vs. nurture.' That is the study of biological as well as sociological factors relating to an individual. From a Layman's point of view, personality means the characteristics and the social skills that an individual has. At the same time, the psychological view of the personality is an explanation about why people with similar heredity, demographics, experience, education, etc., behave differently in similar environmental situations and vice versa.

Even the researchers (see for example, Grzywacz and Carlson (2007)) who defined WLB as a socio-demographic concept, claimed that they are not invalidating the psychological nature of the WLB rather defining WLB from an alternative angle. That is, the psychological aspect of WLB can never be overlooked. Personality is defined as one of the mainstream branches of psychology that studies how an individual behaves and acts in a unique way (Corr and Matthews, 2009). Personalities (personality traits) constitute the subject matter of psychology (Murray, 2008). Therefore, personality traits being one of the important aspects of personal psychology, should have a relationship with WLB as the latter is a construct that has its root in personal psychology.

4.2 PERSONAL PSYCHOLOGY AND WLB

Personal perception about the current life situation in contrast with the 'imagined or planned life' is the one of the fundamental element that of WLB. Personality is an important factor that has a strong impact on an individual and his ability to take decisions. (Allen, Herst, Bruck, and Sttton, 2000) stressed the need for psychometric measures to detect conflicts. WLB and happiness are related to a person's emotion and attitude. The human brain has the ability to work 8, 10, or even 14 hours without any difference in its performance ability (Carnegie, 1986). Banker (2012) identifies the importance of exploring the self in order to keep track with work life balance. Similarly, mastering and controlling self-emotions and feel are the key to achieve balance in life (Gora, 2015). There is an internal domain related to WLB, which is psychological. It is because of this, Malaviya (2012) reported that emotional stability is inversely correlated with an imbalance in the work-life and that internal locus of control has a negative correlation and external locus of control has a positive correlation with work life imbalance. Several studies (Mukesh, 2010; Santosh, 2014; Walia, 2011) have established the emotional intelligence of the employee has a relationship with the employee WLB. Similarly, personality traits have a significant influence in shaping individual attitude, behavior, and perception (Oltmanns and Turkheimer, 2009; Geld, Oosterveld, Heck, and Kuijpers-Jagtman, 2007; Clifton, Turkheimer, and Oltmanns, 2005). That is, the personality traits being a significant influence of individual perception should also have a relationship with WLB as individual perception is one of the fundamental elements of WLB. Therefore, it is concluded that "there is a relationship between personality characteristics processed by the employees and WLB." Hence, the following hypothesis has been formulated:

- *H1: Employee personality characteristics and WLB are related.*
- *H1A: Extraversion and WLB are related.*
- *H1B: Agreeableness and WLB are related.*
- *H1C: Conscientiousness and WLB are related.*
- *H1D: Emotional stability and WLB are related.*
- *H1E: Openness and WLB are related.*

4.3 MATERIALS AND METHODS USED

4.3.1 MEASURES USED

As part of this study, different measures, which are developed by different research scholars as well as academicians for the purpose of measuring certain variables were incorporated in the questionnaire. The following section provides details about such measures, which were used in the questionnaire.

4.3.1.1 WORK-LIFE BALANCE (WLB)

The WLB level of the respondents was measured with the help of a scale for WLB developed by Brough et al. (2014). It is a four itemized five-point agreement rating scale (1 = Strongly Disagree to 5 = Strongly Agree). Out of four items, one item is negatively worded. In addition, the highest score indicates a greater WLB perception of the respondents. At the same time, the lowest score indicates a lower level of WLB perception of the respondents. An indicative item of the WLB inventory was 'I have difficulty in balancing my work and non-work activities.' The score final score of Brough et al. scale of WLB ranges from 1 to 5. Whereas scores from 1 to 2 indicate poor WLB level, scores from 2.1 to 3.9 indicate a moderate level of WLB, and score from 4 to 5 indicate a high level of WLB. The Cronbach's alpha reliability of the WLB scale was 0.864.

4.3.1.2 PERSONALITY

Gosling, Rentfrow, and Swann (2003) developed a very short measure for Big-Five personality, which is known as TIPI, with the objective of using in those situations where the primary objective is not to identify the personality dimension. Here, in order to measure the personality of the respondents' TIPI was used. The TIPI is a ten itemized 7-point agreement measurement scale (1 = Strongly Disagree to 7 Strongly Agree), which measures the five dimensions of the B-5 personality model. Each dimension is measured with the help of two items. Indicative items of the inventory were 'I see myself as extraverted, enthusiastic,' 'I see myself as reserved, quiet,' 'I see myself as conventional, uncreative,' etc.

4.3.2 DATA AND SAMPLE

The study is based on the primary data, which is retrieved from the database created as part of Mr. Joshin Joseph's doctoral degree research work during the period 2015–2019. The database contains 520 sample data obtained randomly (using multi-stage random sampling) from the employees who are employed in the service sector, specifically from hotels, hospitals, and banks. The questionnaire was the tool used to obtain the data, both online (WhatsApp Questionnaire and E-mail Questionnaire) as well as offline (traditional hard copy Questionnaire) questionnaire was used to obtain data. For the purpose of analysis, R version 3.4.3, has been used, and the relationship was identified as well as estimated with the help of general linear modeling (GLM).

4.3.2.1 DEMOGRAPHIC PROFILE OF RESPONDENTS

Out of 520 data obtained, 212 data were obtained from the hospital sector, 177 from the banking sector, and 131 from the hotel sector, respectively. While considering the gender profile, 50.38% (262 no's) of the respondents were male, and 49.62% (258 no's) of the respondents were female. With regard to the education profile, 68.46% of the respondents have an educational qualification equivalent to or above graduation (43.27% were graduates, and 25.19% were postgraduates, respectively). Also, one fifth (20.96%) of the respondents have +2 or equivalent qualifications. Similarly, 10.58% were matriculated.

Similarly, the distribution of the respondents based on income indicates that one fourth (25%) of the respondents reported to have personal income between 20,001 to 30,000 and another one quarter (27.70%) of respondents with personal income between 30,001 to 50,000. It was followed by 22.90% of the respondents with personal income less than 20,000 and 14.20% of the respondents with personal income between 50,000 to 80,000. Table 4.2 further illustrates that one-tenth of the respondents (10.00%) have income above 80,000.

4.4 ANALYSIS AND DISCUSSION

WLB is often defined as a psychological concept based on the perception of the employees (Valcour, 2007; Brough et al., 2014). The objective of this section is to understand the relationship between psychological factors and WLB. That is, to examine the statistical validity of the hypothesized claim that 'there is a relationship between personality characteristics processed by the employees and WLB.' Here in this study, personality is assessed using the B-5 model of personality.

4.4.1 EXTRAVERSION AND WLB

In order to understand the relationship between the WLB level of the employee and extraversion characteristics, WLB is regressed on extraversion. The rules of simple linear regression were scrutinized. Though the data found to be normal and homoscedastic, the scatter plot indicated the absence of a linear relationship between the WLB level of the employee and extraversion (see Figure 4.1). The straight line parallel to the x-axis indicates that there is no linear relationship between the WLB of the employee and the level of extraversion (see Figure 4.1). In the absence of a direct linear relationship between WLB and extraversion, the quadratic effect has been examined. The quadratic relationship is shown in Figure 4.1. The concave line indicates that the relationship between extraversion and WLB is curved positive in nature. In order to examine the relationship between WLB and extraversion statistically. That is, whether the quadratic model between WLB and extraversion is statistically significant in comparison with the linear model of the relationship between WLB and extraversion partial F-test has been used. Table 4.1 illustrates the result of the hierarchical linear regression of WLB on extraversion.

TABLE 4.1 Hierarchical Linear Regression of WLB on Extraversion Using Linear and Quadratic Model

Variables	Model 1			Model 2		
	B	SE B	β	B	SE B	β
Extraversion	0.01	0.02	0.02ns	0.01	0.04	0.05ns
Extraversion ^2				0.03	0.01	0.15**
R^2		0.00			0.02	
Partial F					0.02**	

As shown in Table 4.1, in model 1, illustrates the linear relationship between WLB and extraversion. When the relationship between extraversion and WLB is considered as linear, WLB is not significantly influenced by the employee extraversion level (see Table 4.1, β = 0.02, p > 0.05, R^2 = 0.00). Furthermore, when the quadratic effect is considered (i.e., the model 2), the extraversion has then turned to a significant predictor of WLB (see Table 4.1, Extraversion: β =.15, p < 0.05, R^2 = 0.00; Extraversion^2: β = 0.23, p < 0.01, R^2 = 0.02). Similarly, the quadratic model of extraversion can explain significantly larger amount of variance in WLB in comparison with the explanatory capability of the linear model of extraversion to explain WLB (see Table 4.1, Partial F = 11.22, p < 0.01).

4.4.1.1 DISCUSSION

As illustrated in Figure 4.1, the WLB level of the employee and extraversion level were not linearly associated. On the other hand, the relationship between extraversion and WLB is curved positive in nature. That is, the WLB level diminishes with the increase in the level of extraversion, but after reaching the point of saturation, the WLB level starts to increase together with the increase in the level of extraversion.

Though there is a significant relationship between the extraversion level of the employee and the WLB level, the strength of the relationship between

Exploring the Relationship Between Personality 61

extraversion and WLB is small ($R^2 = 0.02$, see Table 4.1). Furthermore, $R^2 = 0.02$ (see Table 4.1) indicates that the extraversion level of the employee has the potential to explain around 2% of the variance in the level of employee WLB. In the case of regression, R^2 less than .05 is an indication that the relationship between variables was small (Keith, 2015). Therefore, it is concluded that, though there is a significant relationship between extraversion and WLB, the influence that the extraversion can have on WLB is limited (i.e., up to 2% as per the model, $R^2 = 0.02$, see Table 4.1). The finding that there is no relationship between extraversion and WLB is consistent with the finding of Bhalla and Kang (2018).

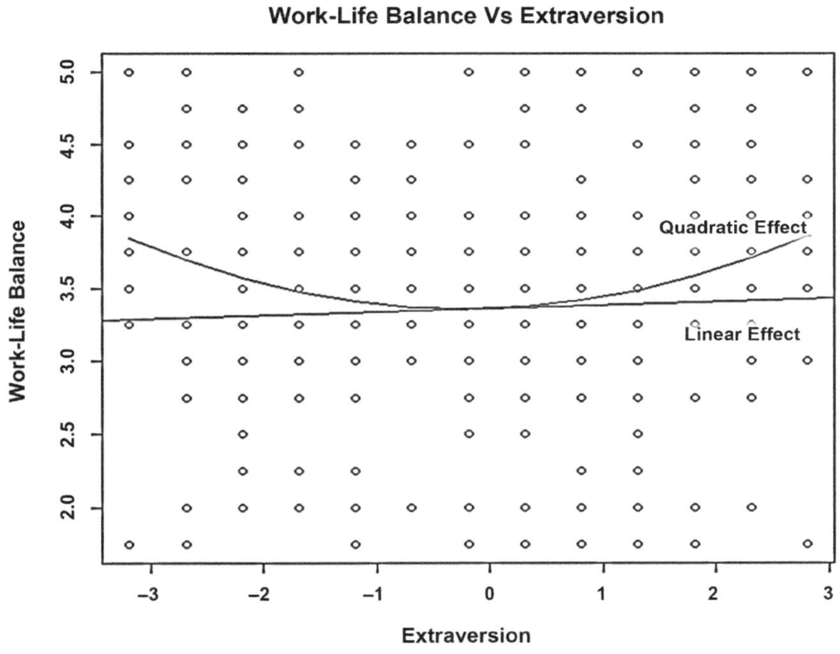

Work-Life Balance Vs Extraversion

FIGURE 4.1 Scatter plot illustrating the relationship between extraversion and WLB.

4.4.2 AGREEABLENESS AND WLB

In order to understand the relationship between agreeableness and the WLB level of the respondents, the WLB level is regressed on agreeableness. Table 4.2 illustrates the summary statistics of the model where WLB is regressed on the agreeableness level of the respondents.

TABLE 4.2 Summary Statistics of Simple Linear Regression of WLB on Employee Agreeableness (N = 520)

Variable	B	SE B	β
Agreeableness	−0.22	0.03	−0.35*
R^2		0.12	
Adjusted R^2		0.12	
F		72.85*	

*a = 3.51.
Source: Primary data.

As shown in Table 4.2, agreeableness has the potential to explain WLB significantly, F (1, 518) = 72.85, p < 0.001, $R^2 = 0.12$. That is, there is a significant relationship between agreeableness and WLB, and hence, the proposed hypothesis is accepted. At the same time, the beta coefficient and standardized beta coefficient were −0.22 and −0.35, respectively.

4.4.2.1 DISCUSSION

As shown in Table 4.2, there is a significant relationship between the agreeableness level of the respondents and the respondent's WLB level. Agreeableness has the potential to explain approximately 12% variance in the employee WLB level ($R^2 = 0.12$). Similarly, the relationship between WLB and agreeableness is inverse in nature (B = −0.22). That is, when the agreeableness level increases, the WLB level of the employee diminishes accordingly. Furthermore, the standardized beta value of −0.35 indicates that an increase in agreeableness by one standard deviation will decrease the WLB level by 0.35 standard deviations. While considering the effect size, agreeableness is a small to medium scale predictor of employee WLB level ($R^2 = 0.12$).

4.4.3 CONSCIENTIOUSNESS AND WLB

For the purpose of understanding the relationship between employee WLB level and conscientiousness level of the employee, the WLB level was regressed on contentiousness. Table 4.3 shows the summary statistics of the simple linear regression of WLB on employee conscientiousness level.

TABLE 4.3 Summary Statistics of Simple Linear Regression of WLB on Employee Conscientiousness (N = 520)

Variable	B	SE B	β
Conscientiousness	0.20	0.03	0.30*
R^2		0.09	
Adjusted R^2		0.09	
F		52.79*	

*$a = 3.51$.
Source: Primary data.

As illustrated in Table 4.3, conscientiousness can explain significant amount of variance in employee level of WLB F (1, 518) = 52.79, p < 0.001, $R^2 = 0.09$. Therefore, it is concluded that there is a significant relationship between employee conscientiousness level and WLB. Hence, the hypothesis that there is a relationship between conscientiousness and WLB is said to be accepted.

4.4.3.1 DISCUSSION

According to the statistics illustrated in Table 4.3, the relationship between WLB and conscientiousness level of the employee is positive in nature (β = 0.30, p < 0.01). That is when the conscientiousness increases, the WLB increases, and vice versa. Similarly, conscientiousness has the potential to explain approximately 9% of the variance in the level of WLB ($R^2 = 0.09$, see Table 4.3). Furthermore, the standardized beta coefficient of 0.30, indicates that an increase in employee conscientiousness by one standard deviation will push up the level of WLB by.34 standard deviations, provided all other factors affecting WLB and conscientiousness are controlled. However, the strength of the relationship between conscientiousness and WLB is small ($R^2 = 0.09$). The finding that there is a positive relationship between WLB and conscientiousness can be considered consistent with the finding of Bhalla and Kang (2018). Bhalla and Kang (2018) found that conscientiousness can enhance work-life facilitation, though it doesn't have the potential to reduce work-life conflict. The finding of Bhalla and Kang (2018) further validates (i.e., conscientiousness can have only limited influence on employee WLB through work-life facilitation) the small effect that conscientiousness has on employee WLB.

4.4.4 EMOTIONAL STABILITY AND WLB

For the purpose of understanding the relationship between WLB and emotional stability, the WLB level was regressed on emotional stability. Table 4.4 illustrates the result of the simple linear regression of WLB on emotional stability.

TABLE 4.4 Summary Statistics of Simple Linear Regression of WLB on Emotional Stability (N = 520)

Variable	B	SE B	β
Emotional stability	0.32	0.03	0.49*
R^2		0.15	
Adjusted R^2		0.15	
F		91.62*	

*$a = 3.51$.
Source: Primary data.

According to the statistics shown in Table 4.4, emotional stability has the potential to explain the employee WLB level significantly F (1, 518) = 91.62, p < 0.001, R^2 = 15. Therefore, the hypothesis that emotional stability and WLB are related is said to be accepted. Hence, it is concluded that there is a significant relationship between emotional stability and the WLB level of the employee.

4.4.4.1 DISCUSSION

According to the statistics shown in Table 4.4, there exists a significant positive relationship between WLB and the emotional stability of the employee (B = 0.32, see Table 4.4). Furthermore, the analysis indicates that the emotional stability of the employee itself has the potential to explain 15% of the variance in the employee WLB level (R^2 = 15, see Table 4.4). Similarly, a standardized beta coefficient of 0.39 represents that an increase in the rate of emotional stability by one standard deviation will result in the enhancement of employee level of WLB level by 0.39 standard deviations, provided all the other factors affecting WLB and emotional stability are controlled.

While considering the effect size of the relationship between the emotional stability and employee WLB level, the emotional stability of the employee has a medium-sized effect on employee WLB level (see Table 4.4, R^2 = 15).

Exploring the Relationship Between Personality 65

Furthermore, the finding that there is a positive relationship between WLB and emotional stability is consistent with the findings of several studies (e.g., Shylaja, and Prasad, 2017; Rajkumar, 2014; Susi, 2014; Malaviya, 2012; Carmeli, 2003).

4.4.5 OPENNESS AND WLB

To understand the relationship between WLB and openness, the WLB level was regressed on employee openness. Table 4.5 illustrates the result of the simple linear regression of WLB on employee openness.

TABLE 4.5 Summary Statistics of Simple Linear Regression Predicting WLB based on Employee Openness (N = 520)

Variable	B	SE B	β
Employee openness	0.31	0.03	0.39*
R^2		0.15	
Adjusted R^2		0.15	
F		89.95*	

*$a = 3.51$.
Source: Primary data.

As per the statistics illustrated in Table 4.5, the openness of the employee explains WLB significantly F (1, 518) = 89.95, p < 0.001, R^2 = 15, indicating that there is a significant relationship between openness and WLB. Therefore, the hypothesis that there is a relationship between openness and WLB is found valid.

4.4.5.1 DISCUSSION

Regressing WLB on employee openness revealed that there is a significant relationship between WLB and openness characteristics of the employee (see Table 4.5). Similarly, the openness of the employee has the potential to explain approximately 15% of the variance in the level of WLB (R^2 = 15, see Table 4.5). The positive beta coefficient is an indication that there is a positive relationship between employee WLB level and openness. Furthermore, the standardized beta coefficient of .39 indicates that an increase in employee openness by one standard deviation will push up the level of WLB by 0.39 units of standard deviations, which can fairly be ascertained as a fair influence, provided all other factors affecting WLB and employee openness are controlled. The effect size indicates that employee openness has a medium effect on employee WLB ($R^2 = 15$). The finding that there is a positive relationship between WLB and employee openness characteristics is consistent with the findings of Bhalla and Kang (2018).

4.4.6 PERSONALITY AND WLB

According to the big-five model, the personality of an individual is identified through five factors viz., extraversion, agreeableness, conscientiousness, emotional stability, and openness. That is, to understand the relationship between personality and WLB, the combined effect relationship between big-five factors and personality need to be assessed. The individualistic analysis of the personality dimensions found that all the five personality dimensions have a statistically significant relationship with the employee WLB level (see Tables 4.1–4.5). However, the personality dimensions are not individualistic in nature; it's a group force and correlated with each other. Therefore, in order to understand the true effect of personality on WLB, all the factors should be considered together rather than considering independently.

In order to understand the relationship between personality and WLB, various dimensions of personality are regressed on WLB, and Table 4.6 illustrates the summary statistics of multiple linear regressions predicting WLB based on personality.

TABLE 4.6 Summary Statistics Multiple Linear Regression of WLB on Personality (N = 520)

Variable	B	SE B	β
Extraversion	0.03	0.02	0.05^{ns}
Agreeableness	−0.12	0.02	−0.20*
Conscientiousness	0.18	0.03	0.26*
Emotional stability	0.18	0.03	0.22*
Openness	0.22	0.03	0.28*
R^2		0.36	
Adjusted R^2		0.35	
F		57.42*	

*$a = 3.51$ (i.e., constant/intercept).
$f^2 = .56$.
Source: Primary data.

As shown in Table 4.6, personality is a significant predictor of WLB $F (5, 514) = 57.42$, $p < 0.001$, $R^2 = .36$. That is, there is a significant relationship between personality and WLB. Therefore, the hypothesis 'There is a significant relationship between personality and WLB' is accepted and concluded that there is a relationship between personality and WLB.

4.4.6.1 DISCUSSION

As per the statistics illustrated in Table 4.6, the five dimensions of personality together can predict approximately 36% of the variance in the WLB level ($R^2 = 0.36$); that is, personality has a large effect on the WLB level. When the value of R^2 is greater than 0.25 for regression, then it is concluded that the predictor has a large effect on the dependent variable (Keith, 2015). When the WLB was regressed on personality factors simultaneously, all the factors were significant predictors of WLB expect extraversion (Extraversion: $\beta = 0.05$, $p > 0.05$; Agreeableness: $\beta = -0.20$, $p > 0.001$; Conscientiousness: $\beta = .26$, $p > 0.001$; Emotional Stability: $\beta = .22$, $p > 0.001$; Openness: $\beta = .28$, $p > 0.001$). That is, the effect of extraversion (extraversion had a significant relationship with WLB when WLB was regressed on extraversion individually, see Table 4.1) on WLB has been mitigated by other factors (dimensions) of personality. It is a clear indication of the nexus of relationships between various dimensions of personality.

Similarly, multiple regression of WLB on personality factors produced an intercept of 3.51 (intercept = 3.51, $p < 0.001$, see Table 4.6), which is the level of WLB at the mean level of personality (the personality scores have been centered prior to regression). Mean personality means the average personality score of the total sample frame. It means that a normal employee (on the basis of personality score on TIPI) has the WLB score of 3.51 on the Brough et al. scale of WLB. A score of 3.51 on the Brough et al. scale of WLB is the state of WLB. Therefore, it can be concluded that an employee with a normal personality always has a balanced work-life. Furthermore, the standardized beta coefficient indicates that out of five personality dimensions, openness has the greatest impact on an employee's WLB level ($\beta = 0.28$, $p < 0.001$), it was followed by the conscientiousness ($\beta = 0.26$, $p < 0.001$), then by emotional stability ($\beta = 0.22$, $p < 0.001$) and by agreeableness ($\beta = -0.20$, $p < 0.001$). Extraversion has the lowest beta value, and its impact on the WLB level is statistically insignificant, too ($\beta = 0.05$, $p > 0.05$).

Taken together, personality characteristics such as conscientiousness, openness, and emotional stability have a positive relationship with WLB.

However, the relationship between agreeableness and WLB is negative. That is, when the agreeableness increases, WLB decreases, and vice versa. However, the extraversion doesn't have any significant relationship with WLB when it is considered together with other personality characteristics. The finding that employee personality has a significant relationship with employee WLB is consistent with the finding of several studies (e.g., Bhalla and Kang, 2018; Mugeanyi, 2017; Moshoeu, 2017; Wu, 2017; Direnzo, Greenhaus, and Weer, 2016; Rantanen, Kinnunen, Mauno, and Tement, 2013).

4.5 CONCLUSION

The findings reveal that employee personality characteristics have the potential to explain the employee level of WLB. Furthermore, personal psychology itself has the potential to explain 32% variations in employee WLB level. It is an indication that employee personality has a large effect in shaping employee WLB. As it has already been validated that the demographics have a relationship with employee personality as well as with employee WLB, the finding of this study exposes the interplay of employee personality characteristics as an intervening variable between employee demographics and personality. However, the relationship with regard to this extends the need to validated further. In this study, personality is measured with TIPI, a shortened version of the big-five model which less robust in comparison with the standardized multi-item measure of personality (see Gosling, Rentfrow, and Swann (2003), for more details about TIPI). Similarly, personality is a construct that is longitudinal in nature; hence the longitudinal design can only expose the change in employee WLB in response to change in employee personality characteristics.

KEYWORDS

- **big-five model**
- **emotional stability**
- **general linear modeling**
- **hierarchical linear regression**
- **openness**
- **work-life balance**

REFERENCES

Banker, A. K., (2012). *The Valmiki Syndrome Finding the WLB*. Noida: Random House India.

Bhalla, A., & Kang, L. S., (2018). *The Role of Personality in Influencing Work Family Balance Experience: A Study of Indian Journals*. Global Business Review. doi: 10.1177/0972150918779157.

Brough, P., Siu, O. L., O'Driscoll, M., & Timmis, C., (2015). Work–family enrichment and satisfaction: The mediating role of self-efficacy and work-life balance. *The International Journal of Human Resource Management*. doi: 10.1080/09585192.2015.1075574.

Brough, P., Timmsb, C., O'Driscollc, M. P., Kalliathd, T., Siue, O. L., Sitf, C., & Log, D., (2014). Work-life balance: A longitudinal evaluation of a new measure across Australia and New Zealand workers. *The International Journal of Human Resource Management, 25*(19), 2724–2744. doi: 10.1080/09585192.2014.899262.

Carmeli, A., (2003). The relationship between emotional intelligence and work attitudes, behavior. *Journal of Managerial Psychology, 18*(8), 788–813. doi: 10.1108/02683940310511881.

Clifton, A., Turkheimer, E., & Oltmanns, T. F., (2005). Self-and peer perspectives on pathological personality traits and interpersonal problems. *Psychol. Assess.*, 121–135.

Corr, P. J., & Matthews, G., (2009). *The Cambridge Handbook of Personality Psychology*. Cambridge: Cambridge University Press.

Direnzo, M. S., Greenhaus, J. H., & Weer, C. H., (2016). Relationship between protean career orientation and work-life balance: A resource perspective. *Journal of Organizational Behavior*, 4–57.

Geld, P. V., Oosterveld, P., Heck, G. V., & Kuijpers-Jagtman, A. M., (2007). Smile attractiveness. *The Angle Orthodontist*, 759–765.

Greenhaus, J. H., & Allen, T. D., (2002). Work-family balances a review and extension of literature. In: Quick, J. C., & Tetrick, L. E., (eds.), *Hand Book of Occupational Health and Psychology* (pp. 165–183). Washington: American Psychological Association.

Grzywacz, J., & Carlson, D. S., (2007). Conceptualizing work-family balance: Implications for practice and research. *Advances in Developing Human Resources, 9*(4), 455–471. doi: 10.1177/1523422307305487.

Hilla, E. J., Grzywaczb, J. G., Allena, S., Blancharda, V. L., Matz-Costac, C., Shulkinc, S., & Pitt-Catsouphesc, M., (2008). Defining and conceptualizing workplace flexibility. *Community, Work and Family*, 149–163. doi: 10.1080/13668800802024678.

Hutton, E. L., (1945). What is meant by personality? *The British Journal of Psychiatry*, 153–165.

Kalliath, T., & Brough, P., (2008). Work-life balance: A review of the meaning of the balance construct. *Journal of Management and Organization, 14*(3), 323–327.

Keith, T. Z., (2015). *Multiple Regression and Beyond*. New York: Taylor & Francis.

Kopelman, R. E., Greenhaus, J., & Connolly, T. F., (1983). A model of work, family, and inter-role conflict: A construct validation study. *Organizational Behavior and Human Performance*, 198–215.

Moshoeu, A. N., (2017). *A Model of Personality Traits and Work-Life Balance as Determinants of Employee Engagement*. University of South Africa. Abigail Ngokwana Moshoeu.

Mugeanyi, N., (2017). *Work Life Balance: The Need for Self-Awareness and Care*. Retrieved from: Researchgate.net: https://www.researchgate.net/publication/318910632 (accessed on 21 October 2020).

Mukhtar, F., (2012). *Work Life Balance and Job Satisfaction among Faculty at Iowa State University*. Iowa State University. Iowa: Farah Mukhtar.

Murray, A. H., (2008). Chapter 1: Introduction. In: Murray, H. A., (ed.), *Explorations in Personality* (pp. 3–35). New York: Oxford University Press.

Oltmanns, T. F., & Turkheimer, E., (2009). Person perception and personality pathology. *Curr. Dir. Psychol. Sc.*, 32–36.

Rajkumar, R., (2014). *Work Life Balance of IT Professionals in Relation to their Self Concept, Hardiness, and Emotional Maturity*. PhD Thesis, Annamalai University, Department of Business Administration, Annamalai Nagar.

Rantanen, J., Kinnunen, U., Mauno, S., & Tement, S., (2013). Patterns of conflict and enrichment in work-family balance: A three-dimensional typology. *An International Journal of Work, Health and Organizations, 27*(2), 141–163. doi: 10.1080/02678373.2013.791074.

Shylaja, P., & Prasad, C., (2017). Emotional intelligence and work life balance. *Journal of Business and Management*, 18–21.

Sociology Guide.Com. (2016). *Personality*. Retrieved from: Sociology Guide.

Susi, S., (2014). *A Study on Work-Life Balance among ITES in the Bangalore City*. PhD Thesis, Bharathiar University, Management, Coimbatore.

Timmis, C., Brough, P., Siu, O. L., O'Driscoll, M., & Kalliath, T., (2015). *Handbook of Research on Work-Life Balance in Asia* (pp. 294–314). In cross-cultural impact of work-life balance on health and work outcomes.

Valcour, M., (2007). Work-based resources as moderators of the relationship between work hours and satisfaction with work-family balance. *Journal of Applied Psychology, 92*(6), 1512–1523. doi: 10.1037/0021-9010.92.6.1512.

Voydanoff, P., (2005). Toward a conceptualization of perceived work-family fit and balance: A demands and resources approach. *Journal of Marriage and Family*, 822–836.

Wu, A. H., (2017). *Work-Life Balance: A Study of Personality Factors as a Predictor of Work-Life Boundary Permeability and Use of Enterprise Social Media and Technology*. East Carolina University, East Carolina: Allison H Wu.

CHAPTER 5

Artificial Intelligence as Disruptive Technology: A Boon or a Bane in the Global Business Scenario

SEEMA SAHAI[1] and SAURAV LALL[2]

[1]Associate Professor, Amity International Business School, Amity University, Noida, Uttar Pradesh, India

[2]Azure IoT, Microsoft Seattle, Washington, United States of America

ABSTRACT

The technological revolution, which is taking place around us, has created an apprehension amongst people for fear of loss of jobs and increasing inequality amongst them. It has been said that Artificial Intelligence (AI, hereafter) is a reason for generating this fear. A comparison of past automation and the current scenario shows that AI has reduced human efforts and has also helped in increasing productivity. The influence of AI can be seen in every field in life but do we understand where to stop?

This chapter gives an optimistic approach towards the potentials of AI and the risks involved in the workplace. It includes the following:

- Relating the earlier periods of technology changes with modern evidence on the effects of artificial intelligence.
- Present claims on the effects of artificial intelligence on the future of workers at the workplace.
- Analyzing the fears and claims around artificial intelligence and answer the questions of why and how.

We will further see whether how AI has the ability to increase human productivity and intelligence with collaborating with humans and further help in increasing efficiency at the workplace.

5.1 INTRODUCTION

There has been such rapid growth of technology, especially in artificial intelligence (AI), that people have not been able to assess the impact of it. Statements that say that robots are replacing humans have brought in fear amongst the workers. Organizations are adopting this new technology to remain in the competition. Due to all such advancements, AI has been termed as "disruptive technology."

We can take an example here and explain the impact of AI. Should the organizations go ahead and make way for compensation or severance pay of the staff, or should the organizations estimate the expenditure for investing in robots? Will human resources have to change their name as they will also have to look after the "not-so-human resources"? How many humans would still be working at functional levels?

Looking at the above statements, we can say that it has been forgotten that with all its advantages of being more efficient and faster, AI lacks in emotion. Wherever, creativity and thought are needed, no machine can replace it.

This chapter will elaborate on the level of awareness, fears, and effects of AI and about the current understanding of AI amongst the employees at the workplace. There is also an aim of identifying positive potentials of AI and where it can be helpful at the workplace. Further, it aims at finding a balance wherein humans and machines can function with collaboration and not replace each other. Its influence in various fields has been discussed. AI has been defined here as the computer systems that can adapt and learn and reflect human-like behavior.

Yes, AI is growing rapidly and affecting various sectors. However, the effect is not equal. Only a few sectors are enjoying the benefits of AI; the rest are not. With the size of technology shrinking rapidly, as seen in the case of computers from being one huge room size computers to desktops to the coming up of laptops, palmtops, and the handheld devices more powerful than a computer. Cloud computing, for instance, is the most efficient and cost-effective storage of data over the cloud. (Brynjolfsson, McAfee, and Manyika, 2014) There has been a shift in the business models (BMs) of today with small, medium-tech startup companies coming up, which are crowdfunded are also contributing to the technological growth of the world.

AI may be a factor for job loss, but its contribution to the field of medicine is a lifesaver for many (Grosz et al., 2016). If we do not evolve ourselves and prepare for disruptive technologies like AI, we may fade away with time and planet ruled by robots and machines (Grosz et al., 2016). There has been an increase in research in the field of AI at a global level hence motivating education institutes and countries channeling their energies in understanding the deep potential and impacts of AI in the modern business world and especially at the workplace. There is a global consensus on the fact that machines created by humans have always enhanced productivity and contributed and improved quality of life.

5.2 UNDERSTANDING THE TIMELINE OF ARTIFICIAL INTELLIGENCE (AI)

The need for AI began almost 70 years back, with the concept that machines can also replicate humans. This idea may seem to be out of fiction movies back then and next to impossible, but today it is a part of our lives and all around us. The time has come when humans are actually feeling threatened by the machines.

5.2.1 THE PERIOD OF DARKNESS-1943

World War II was a platform where many people came together to evolve such technologies and machines, which could help them in becoming victorious in the war. The Mathematician Alan Turning and neurologist Grey Walter took the challenge and decided to build an intelligent machine, which could replicate the intelligence of humans. Inspired by a turtle, they built the first robot which could move. This became a symbol of innovation at that time.

5.2.2 TOP-DOWN APPROACH 1956

The term AI was first coined at Dartmoth University. There were two schools of thought. One stated a top-down approach to use AI that is from a system programmed in advance with set rules that would display human-like behavior. The other school of thought stated a bottom-up approach for AI, for example, brain cells and neural networks that would acquire new learnings over a period of time.

5.2.3 YEAR 1970

Marvin Minsky claimed that in the next four to eight years, there would be a machine with the intelligence of an average human being.

5.2.4 1973

Millions of dollars had been spent on AI developments and research, but the research had shown no promise for the future. Funds for this industry were cut to half because it showed no promise for future growth.

5.2.5 COMMERCIAL VALUE OF AI-1981

The dilemmas for the AI industry got solved by the end of the year when businesses started to realize that the commercial value of AI had started to attract more and more investments.

5.2.6 YEAR 1997-THE DEEP BLUE

The followers of the top-down approach had one last man standing that was Deep blue–a supercomputer. This computer took on the world chess championship. This system could evaluate up to 100 million possible positions in a second. This was far superior to a human brain. Deep blue won the chess championship, and this was a turning point for all AI technologies and the beginning of a legacy.

5.2.7 VENTURING INTO THE SPACE-2001

The possibility of the existence of AI has led to many fiction stories and movies based on it. The movie "Space Odyssey" mentions a system HAL 9000, which is said to be error-free and very accurate. The film showed some of the predictions and assumptions surrounding AI capabilities. What was then considered fiction had become a reality today? What was unimaginable those days is a way of life now.

Artificial Intelligence as Disruptive Technology 75

5.2.8 THE NEW ERA-2002-FIRST ROBOT TO BE USED AT HOME

iRobot Corp. designed the first robot for homes that acted like a vacuum cleaner. This robot was small and efficient in cleaning homes. It had a sensor technology to detect dust and clean the same.

5.2.9 MACHINES FOR WARS-2005

US military wanted to exploit AI technology and strengthen its military power (Crawford, West, and Whittaker, 2019). Therefore, the US military developed a fighter dog known as BigDog. It could take on rough terrains, had 50 sensors, and was very quiet. It also could be used as a bomb disposal robot.

5.2.10 AI CRACKING BIGGER PROBLEMS-2008

The Apple iPhone showed few features of AI technologies with its speech recognition application. It looked very simple feature, but many years of work was put in, and still refinement of the same is going on. It recognizes speech up to 80% accuracy.

5.2.11 DANCING ROBOT-2010

AI led to a shrinking in the size of technologies; now, smaller-packed robots could perform efficiently. Humanoid robots were developed which could do activities that old robots couldn't do. It could even dance in a synchronized way to music (Upchurch, 2018).

5.2.12 MAN VS. MACHINE-2011

On a quiz show, IBM designed a robot that took on human brains. It was far difficult than a game of chess. In the quiz, the IBM robot Watson had to answer riddles and questions with complexities. Watson, the robot, won the quiz show.

5.2.13 MACHINES ARE INTELLIGENT? – 2014

Stories of fiction writers were becoming true from Google's and Tesla's driverless cars to the Skype launch of translation during real-time Skype calls.

5.3 ARTIFICIAL INTELLIGENCE (AI) IN HEALTH CARE

Artificial intelligence in the healthcare sector is life-transforming and promising for the future. It has contributed in drug development, cares at hospitals, and clinic researches. Many countries are spending a huge amount of money on the development of AI for the healthcare sector. The precision in surgery to predicting diagnostics, AI as a tool, works well in collaboration with doctors. AI applications would solve the problem of effectiveness, affordability, and access to healthcare. AI contains the capability to cut costs yet improve the treatment and improve the accessibility to healthcare.

AI will continue to add value to the Healthcare sector and improve the standards of healthcare in the world. However, will AI benefits reach the developing countries, and underdeveloped nations are the question that is unanswered because these countries lack basic healthcare facilities.

The issue with AI use in healthcare is the lack of trust of doctors in AI-enabled technologies. Many surveys have shown that patients are skeptical about trusting their lives in the hands of a robot. Scaling the AI technologies in healthcare is also a problem as it would involve costs. Small and regional healthcare companies may not be in a position to incur huge costs in terms of AI technologies.

5.4 AI SHAPING THE FUTURE

With the help of AI, prosthetic technology has seen much advancement. In the past, one could never have imagined that if one loses an arm or a leg, it could ever be replaced. However, AI has changed the entire scenario in the healthcare sector. Healthcare has evolved over a period of time; earlier injection, pills, and operations would save us from death. The average life expectancy since the early 18's has jumped between 25 and 40 years to approx. 70 years worldwide.

The modern healthcare system is seeing AI evolving and has seen that healthcare technology manages itself. AI technology now can take a patient blood sample and run thousands of tests to do the most efficient diagnostics.

Now, AI technology can predict your healthcare situation, depending on your lifestyle.

No one had ever imagined that the technology which was considered to be a curse is a blessing in disguise, and it is changing the scenario of the healthcare sector. Apple recently has launched Apple Watch that can perform ECG on the patient and monitor his heart ratings. With the advent of AI technology, patients' safety can be ensured every time as AI devices would track the patient's health, and if anything goes wrong, the emergency services would automatically be informed, and there would be a certain chance of survival of the patient. AI technology can also be used to monitor the dosage levels of the patient and prevent from overdosage. AI devices can design the dosage schedule and plan the dosage cycle for patients ensuring that they do not miss any dose and health is maintained.

AI technology has also developed Sleep beds and sleep devices that would prevent sleep disorders, which are becoming very common these days, especially in the US (Agrawal, Gans, and Goldfarb, 2017). The bed uses machine learning and tracks sleeping habits, and provides a sleeping score to the individual.

5.5 ARTIFICIAL INTELLIGENCE (AI) IN MARKETING

Marketing uses data from AI machines to understand customers' needs and wants and improve customer experience through machine learning. Today we are surrounded by huge chunks of data. The world is data-driven, and this big data could be analyzed and managed through machine learning and AI technology.

5.6 POLARIZATION OF JOBS

Rapid automation and robotics are some of the major concerns for the polarization of jobs around the world. Middle skilled and lower-skilled jobs are easily replaced through AI-enabled technologies, thus causing a huge pool of skilled but unemployed workforce in the world.

New AI technologies lead to the substitution of the task performed by current employees to more efficient robots who can do the same task in lesser time and with more productivity and lesser cost to the company. At times AI technologies require human support and care to run. This means that if AI technologies are included in the workplace in some cases, it may increase the number of people employed to ensure the proper functioning of the technologies. The economy would take a toll because if people will not be employed and will not have disposable income, thus will not invest money back into the economy.

How technology change impacts unemployment? Is that when a certain job is replaced by AI, the job would not be visible, and any other chance of future employment would be affected by the same.

Highly automated companies would contribute less to the hiring market; thus, it would lead to a further increase in the level of unemployment. Who will buy the products produced by machines if they do not have the income to buy those products and services, which would lead to the polarization of jobs in the world? With the coming up of AI technologies, jobs have become a smaller task for the machines, and these tasks, once automated, would become a skill for the machine.

Jobs vary from country to country. For the US, if a certain job is a task, it may be a job for India. Thus, jobs vary from cross-cultural and the level of income of the countries. Every country would react to the level of intervention of AI-enabled technologies and may adapt to the same in a different way. A developed country may positively welcome new AI technology. A developing country may be hesitant in bringing technology. Whereas underdeveloped nations may not be in a position to afford AI technology as they lack basic technologies in their country.

However, there are evidences that a machine and human can work in collaboration and help in improving efficiency at the workplace and benefit each other equally without being a threat to humans. If a job becomes a task because of AI's machine learning, rather than killing a job opportunity, man can be hired to see that a machine runs properly and completes a task on time without any limitations.

This example can be seen in the case of pilots flying planes. With the advancement in AI technologies, the plane flies automatically on its own defined path using various algorithms and AI machine learning but still, why we need a pilot to be present in the cockpit to fly the plane. The answer is that technology and humans can work in collaboration, further improving the efficiency and improving the standard of living of people.

Doctors, for instance, while treating a patient, can give each patient more time because his time is saved through AI-run diagnostics run on patients to understand the disease patterns, and quicker healing can be worked on by the doctor.

The irony is that you would need humans to develop technology, then technology can perform human-like activities. The supply of laborers who are skilled and against the ones who are unskilled increased; therefore price of the skilled workforce fell drastically, thus making the hiring of the skilled workforce more viable rather than developing new technology. The installation of complex and bigger artificially run technologies has increased over a period of time, and so has the need for a workforce capable to handle such technologies. It is assumed to be a labor-saving technology (Economics for the Royal Society and British Academy, 2018) (Figure 5.1).

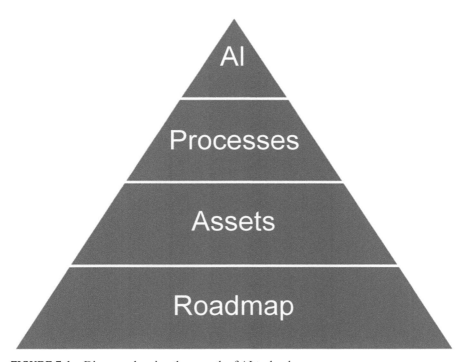

FIGURE 5.1 Diagram showing the growth of AI technology.

The human mind has a limited capacity, whereas AI technology has the capability to perform the same task better than the human mind. The major task of AI technology is to offer expertise and knowledge to people who do not specialize in that area. As doctors can get active feedback from AI technology hence improve the diagnosis of the patients (Figure 5.2).

In the manufacturing industry, AI can remove wastage of resources and increases the productivity of manufacturing. With automation, the defects can be reduced, and production can be enhanced to manifolds. Rather than hiring high skilled workers or medium-skilled professionals, manufacturing could afford hiring technology than hiring a skilled workforce. There has been an increasing consciousness on AI technologies in public spaces talks of fear of AI taking over our jobs and replacing us in the future. A survey in 2016 proved that 6% of people feel fear of AI taking over their jobs and leaving them unemployed.

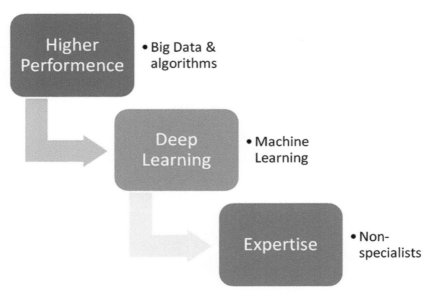

FIGURE 5.2 Diagram showing how non-specialists in a particular area gain expertise through AI technology.

5.7 JOB LOSSES AND GAINS FROM ARTIFICIAL INTELLIGENCE (AI)

Studies have indicated that 15–20% of jobs in England could be automated in the near future. However, some jobs have elements that cannot be automated; hence they require humans in the process. Thus machines and men work in collaboration. The scale of automation is an important question as if the scale is low, there is nothing to be feared, and humans can adapt to that fast-paced technological changes, but if the scale of high, then it is hard for humans to adapt to the technological changes. Jobs which require low skills are at the maximum risk because AI-enabled technologies have the capability to perform complex tasks, and the basic low-skilled job can be easily performed by the machines of this generation (Osoba and Welser, 2017) (Figure 5.3).

Artificial Intelligence as Disruptive Technology

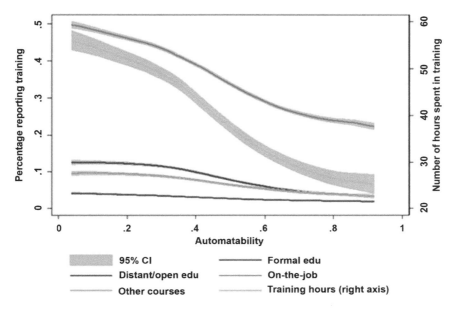

FIGURE 5.3 Training incidence by the degree of job automatability.
Note: Sample includes adults 25–54 years old from 32 countries.
Source: OECD, 2016.

5.8 RECENT DEVELOPMENTS IN ARTIFICIAL INTELLIGENCE (AI)

Robotics is adapting to human behaviors and learning from them. Robots have started learning through observations and machine learning. The robot Nividia has the ability to observe the task performed by different humans. There might be fear that humans will lose to robots one day but have fear when you can collaborate and learn from each other in the same environment rather than competing against each other. If humans could do everything, there would never have been any technology. Technology is for us, not us for the technology, so if we collaborate with technology, we can enjoy its benefits and learn from machine learning (Figure 5.4).

The development of AI involves various steps before its final evolution. The first stage of AI is Machine Learning, which with a study on neural networks and topography of humans with collaboration to computing, leads to the narrowing of AI further, which gives us the artificially intelligent machine.

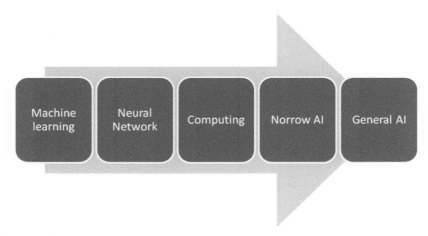

FIGURE 5.4 Diagram shows the pathway to development of AI.

5.9 CYBERSECURITY USING AI

As technology is evolving such rapid space, it needs to be protected from cybercriminals, and here AI comes into place. Cybersecurity through AI technology would detect any activity beforehand only and prevent from big damage. Google developed a cybersecurity company, chronicles, which is a leading cybersecurity company and has the best results in threat detection and prevention of cybercrimes.

5.10 X-RAY BY AI

Healthcare has witnessed AI technologies at a rampant pace and is still evolving in the healthcare sector. AI-enabled technologies help in early diagnosis and precision diagnosis of complicated patient problems and provide assistance to doctors at all levels of patient treatment.

5.11 SMARTPHONE APPS-AI IN OUR HANDS

Our smartphones have become our lives nowadays. In addition, indirectly, so has AI. Research by Gartner shows that 85% of our mobiles will be equipped with AI-enabled technologies as compared to the 8% that are already enabled with AI technologies.

Smartphones use AI technologies like Google Assistant and Siri, which work on voice inputs and does actions on behalf of us and reduces human efforts even further. Suppose you are driving, and you want to send a message just by directing Siri or Google Assistant to do the same.

5.12 FINTECH USING AI ENABLED TECHNOLOGIES

The banking sector has witnessed the AI influence the most, from people visiting the brick and mortar model to now just bank in your pocket on the go. AI has enabled banks to reduce the processing time for cheques. The AI-enabled technology would read the cheque and deposit the same without any manual entry required. Hence replacing the need for a human bank manager to do the same. AI is also protecting the banking sector from cyber crimes and detecting any banking frauds. Usage of chatbots are servicing the clients 24×7 without any requirement of client servicing individual.

5.13 PLATFORMS TO PROVIDE DEEPER LEARNING-AI-BASED

Usage of neural networks to enhance the intelligence of a machine. By understanding the layers of human brains and mirroring the same on a machine. Deeper learning would enhance machine learning hence improve the AI systems, further making it more reliable and friendly.

5.14 BIOMETRICS USING AI

After the thumbprints used to unlock your phone and earlier retina locks, the talk of the town became the face recognition unlocking, which is a product of innovations in AI and machine learning.

5.15 CREATION OF NEW CONTENT USING AI TECHNOLOGIES

Content like videos, ads infographics created by humans can be imitated in a better way by AI-enabled technologies. Further, making content creation cost-effective and less time consuming than when created by the human brain.

5.16 RECOGNIZING EMOTIONS USING AI TECHNOLOGIES

Now the technology has the capability to read human-like emotions and process the same to produce content that would make you feel better in the future. Apart from voice and physical attributes, the machine would have the capability to analyze your attitude, positive or negative, and provide some counseling and mental wellbeing.

5.17 SELF-DRIVING CARS A GIFT OF AI TO HUMANS

Tesla and Google, two big companies, have developed self-driving cars, which have reduced the human efforts to manifold and have provided the ease of transportation and movement of men.

5.18 IMPROVEMENT IN QUANTITY AND QUALITY OF TASKS

AI has made our lives easier and works more fun. It has always been observed since ages that humans were always resistant to change and skeptical about every invention and creation at the first place and over a period of time adapted the same. Similar is the case with AI technologies initially, there are debates and discussions about the negative impact of AI-enabled technologies, but we are unaware that we are surrounded by the same at all times, and we think its basic technological device.

India being a service-focused country, should be aware of how AI-enabled technologies could impact its job market, and the increasing population for both India and China should be an alert for the trouble they may face if they do not adapt to the technologies and work on the skill development (Figure 5.5).

AI can only be used to the maximum with human involvement in its birth and development. AI-enabled technologies not only make our lives easier, but they also solve the complex problems quickly and effectively.

5.19 LARGE SCALE IMPACT OF AI ON OUR LIVES

Whether an AI technology replaces or compliments human for a task still, it is an extension of human capabilities. AI provides various solutions to our economic and social problems. The economic benefits and wage increases took time to emerge, and major displacements of people took place in the process (The British Academy and The Royal Society, 2018). The world has seen a reduction in friction between humans after the advent of AI developed technologies.

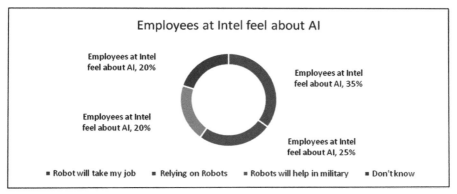

FIGURE 5.5 Pie chart showing the fear and thoughts surrounding AI technologies.
Source: Intel (2017).

Human is considered to be the most important resource for a company. Therefore, AI technology is used in Human resource analytics to analyze the employees' behavior and how they react to different changes in the work environment (Kolbjornsrud, Amico, and Thomas, 2016).

5.20 INVESTMENTS IN DIGITAL INFRASTRUCTURE FOR AI-ENABLED TECHNOLOGIES

Many countries worldwide are investing in AI technologies, and there are various startups coming up in the AI industry. China and the USA are the two countries which dominate in startup funding are in the AI technology sector.

If India wants to compete with the likes of the US and China, thus it should focus on developing skills and promote the coming up of startups and newer innovation. Funds should be directed towards research and development, also motivating young entrepreneurs to start their own business and work towards digitization of the economy.

In recent times, India has seen major developments in the science and technology sectors. However, these are just the baby steps.

If India wants to be the fastest developing economy, then it has to focus on research and development and increase the skilled workforce to compete with the world. India also faces a major challenge of brain drain. That is, young Indians who are skilled get better opportunities abroad, and this leads to the drain of India's wealth. If India needs to prevent this damage, then it has to provide better opportunities for growth enhanced standards of living for its citizens. If AI-enabled technologies enter India at a rampant pace, many Indians may lose jobs because of the lack of skill and expertise required to handle the same.

5.21 ARTIFICIAL INTELLIGENCE (AI) USED IN VARIOUS SECTORS

The pie chart shows the usage of AI technologies in various sectors. AI-enabled technologies are used most in Healthcare, cybersecurity, finance, and sales and marketing. Sectors like analytics, the internet of things (IoT), commerce, and personal assistance is slowly growing in size and use of AI technologies.

It is important to understand the sector's wise usage of AI so that the unequal use in the other sectors is covered through research and development for that sector and investments in those industries. Overdependence on technology is also not good for any sector as it may lead to job losses and people losing faith in technology. Humans and technology should go hand in hand and not in different directions. If humans go on to use natural resources and harming nature, there would be a day when nothing would be left for our grandchildren. Similarly, if humans would stop developing technologies because of fear of job losses and loss of their livelihood, our next generations could never enjoy the advancements in the technologies that we enjoy around us today.

If we want to use the most of the AI technologies around us, we must accept the technologies with open hands and do not fear from use of artificially intelligent technology; rather, should try to make the most of it available and learn from the machine learning and new AI technology (Figure 5.6).

Artificial Intelligence as Disruptive Technology

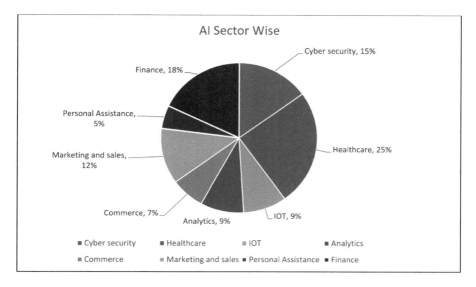

FIGURE 5.6 Pie chart showing the sector-wise use of AI technologies.
Source: CB Insights (2017).

5.22 BOON OR BANE?

The question that comes to our minds is that AI is becoming a buzzword all around us. AI is the future of all innovations and research. A lot of organizations have realized the value of AI and how they can enhance their productivity and use the most out of human capabilities well matched with the AI-enabled technologies.

It's all about gaining the first mover's advantage. The organizations that would realize the true potential of AI-enabled technologies would create disruption in the industry, whereas organizations who fail to adapt with the changing technology and exploiting the benefits of AI would perish with time (Figure 5.7).

5.23 HOPE FOR FUTURE

Many organizations feel that AI may not lead to job losses; rather, it holds more job opportunities in future. With more systems being installed at workplaces, the need for humans handling those machines would be needed more. Yes, AI

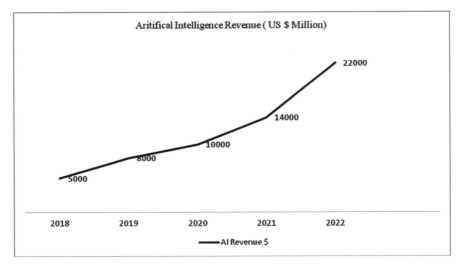

FIGURE 5.7 Graph showing artificial intelligence revenue for the world market.
Source: Tractica (2016).

may replace repetitive and monotonous jobs yet not risk the loss of jobs for many as if one door closes, a hundred new doors would be open for us.

AI is the future for humans. With the constant skill upgradation and knowledge sharing, AI-enabled technologies can be used with the human brains, which would help further in developing new technologies and making our lives easier.

5.24 ADVANTAGES OF AI AT THE WORKPLACE

Every business is hesitant in acquiring new AI technologies. There are trust issues and preconceived notions about AI-enabled technologies. If the mind is clear from all assumptions about AI humans, can exploit the most out of AI-enabled technologies and make the most of it.

Another debate is the inequality of AI technologies for various parts of the societies (Gries and Naudé, 2005). However, governments have realized the importance of AI and how it can benefit humans in increasing our productivity, reducing our costs in terms of time and money. If AI-enabled technologies are not used in businesses, efficiency in operations may not be achieved. It is important for every business to upgrade the skills of its employees on a regular basis to adapt to the changing technological environment. Every individual is scared of losing his or her daily bread to a machine. However, what we forget is that machines are used to help us increase our efficiency and improve our standards of living.

We can improve the education level to ensure individuals are prepared well in advance to accept the changing technological environment and accept the world of humans with AI-enabled technologies.

5.25 POLITICAL ENVIRONMENT IMPACTING ARTIFICIAL INTELLIGENCE (AI) ENABLED TECHNOLOGIES

In earlier times, taxes applied on AI-enabled technologies was less, so it promoted research and development in AI technologies and companies earning profits. However, because of the inequality in society, the government decided to levy more taxes on AI-enabled technologies. Many developing countries have put various restrictions and regulations on AI-enabled technologies because of various health hazards and to protect the interest of their citizens.

The political scenario in every country is very volatile. With the government changing regularly in many countries, the AI industry fails to extract the most benefits out of the existing government policies and regulations. Many NGOs also protect the interest of society and restricts the research and developments of many AI researches all over the world.

If the governments do not provide the support to the AI industry, the businesses will gradually move out and hence hinder the technological growth of the world. With the advent of artificially intelligent technology, the youth especially are most attracted towards the field of science and technology. The world is getting more engineers and scientists now than from the past. If the youth is interested in the technological growth provided by AI technologies, there would be a constant upgradation of skills and more equality of AI development through various segments of the industry. Governments play a major role in implementing such policies that would benefit AI businesses and hence improving the standard of living of individuals who can enjoy the benefits of AI technologies.

5.26 ECONOMIC ENVIRONMENT SURROUNDING THE ARTIFICIAL INTELLIGENCE (AI) SECTOR

With India becoming the fastest growing economies in the world, India is seeing and an upward trend in startups and new business ideas in the field of science and technology. India is the IT hub of the world. Famous for its availability of cheap and skilled workforce. India is focusing on skill upgradation and digitization of its economy.

With the upcoming industrial boom and technological growth in India, the AI business can gain great benefits from its economic background and huge market potential. With the infrastructure having a boom in India, AI technologies would gain an advantage in the Indian market.

The disposable income of individuals in India is increasing as the standards of living are improving. As the standards of living are improving, so is the need for newer advanced technologies increasing (Figure 5.8).

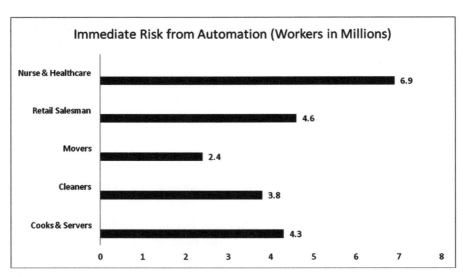

FIGURE 5.8 Graph showing jobs which have immediate risk from automation by 2025.
Source: CB Insights (2016).

Humans need to take the matter into their own hands. If automation would replace our jobs, then how would humans survive? (Kolb and Jones, 2012). The answer lies within us all; if humans continue to take things easy and lightly their might be a time when humanoids or robots would rule our planet, and we would become extinct and stagnated. The only way is to collaborate with technology; if humans could upgrade their skills constantly and try and understand the importance of technologies and how technologies can shape the future, then humans would never run out of jobs. The key to survive is to collaborate with the machines. If technologies would become human-like, then what would be left to differentiate us from that of a robot.

5.27 SOCIETAL FACTORS THAT WOULD AFFECT AI TECHNOLOGIES IN THE WORLD

Society plays an integral part in developing an individual's attitude and personality. Society shapes our thinking patterns and our responses to life situations and problems. Humans are feared of losing jobs, death, fear of no money, fear of no shelter. This fear is induced by the norms of society. Initially, during the older times, man uses to live in caves and in the deserts without complaining about the same. However, civilizations have grown over a period of time.

Every culture in the world has a different meaning; for example, number 4 is considered to be bad in china. White color clothes at the wedding are considered death in Indian weddings whereas in western countries the bride wears white at the wedding. With such varied cultures and societal norms, how would AI get universal applicability and acceptance? Every society would take the same technology in a different manner. For example, self-driving cars are considered to be best for US markets, but in India, self-driving cars are considered to be a risk to the jobs of hundreds of cab drivers.

So when we talk about equality of technologies around the world. It becomes difficult to get an equal stature at every level of society. Every nation would accept AI technologies in a different manner. Some would be open for 100% of AI-enabled technologies, whereas some just 50%. As it's a common fact that change is the only constant. The world around us is constantly evolving, and it's all about how quickly we adapt to these changes and move forward in a dynamic yet predictable environment. Why we say predictable because we know that if humans would not work on skills upgradation and not work on a collaboration with the technologies, we might get replaced by them soon.

5.28 EXPONENTIAL CHANGE

The world is facing an exponential change, and so are the technologies. However, are we ready for this exponential change that is happening around us? AI is shaping the way humans are leading their lives. If humans and technologies do not collaborate, then efficiency could not be achieved. AI is the future today. With the AI technologies entering into the healthcare and education sectors, the world is seeing hope for a brighter future (Jarrahi, 2018).

In the healthcare sector, AI technologies have helped in increasing the life of patients all over the world. Technologies like pacemakers and prosthetics are giving a second chance to patients today. Many fear that one-day AI technologies might replace doctors, and it could lead to putting the lives of thousands in the hands of a machine. AI technologies aim at providing technologies that could think and behave like humans. With the advent of sophisticated technologies, the skills required to understand and decode such technologies has also increased exponentially. If AI technologies had not been developed, we could never have enjoyed the services like making payments online just sitting at home or having your bank accounts on the go in your smartphones.

With the passage of time, the technologies are shrinking in shape and sizes, and a small chip can control and function as the heart. With such AI technological advancements, humans need constant upgradation of skill and more investments in research and development. We are in the era of talking and walking machines that could handle complex problems and display human-like behaviors. The question arises that can machines be trusted? A machine is a preprogrammed computer application that would perform the assigned task in the manner it is programmed to do in. but when it comes to making a decision according to the dynamic changing environment according to the situation, a robot may fail in that case. A robot can display human-like behavior, but it cannot be like humans.

If humans would understand the importance of AI-enabled technologies and make the most use of it to perform the same task more efficiently and within less time than machines are a boon for humans. AI technologies should be promoted from the school level so that we prepare our future generations for the challenges and opportunities that the future holds in terms of developments in AI-enabled technologies. If students would understand the importance of AI-enabled technologies, more skill-sets could be developed, and newer, better technologies could be developed to help make the world a better place to live.

5.29 CONCLUSION

It is apparent that AI is penetrating every aspect of our life. Some have mentioned it as a disruptive technology. However, it is still to be seen whether this disruptive technology is proving to be a boon for us or a bane. "While no one knows what AI's effect on work will be, we can all agree on one thing: it's disruptive" (Wired Insider, n.d.). As many researchers have pointed out, there are more positive aspects to it than negative. In all aspects, AI brings about changes and precision that a human being cannot assure at all times. Starting from simple calculation to greater tasks like reaching to the moon and other planets is done in a more precise manner than when done without these devices. The human brain is involved totally in building such an artificial intelligent device. What we fear is that will our own making start ruling us? Will this technology become so powerful that it will make us its slave? These questions can only be answered in the future where we will see exactly what is happening. However, so far, this technology is being used to its maximum capacity to make a better life for human beings.

KEYWORDS

- **AI-enabled technologies**
- **artificial intelligence**
- **deep learning**
- **digital infrastructure**
- **disruptive technology**
- **startup funding**

REFERENCES

Agrawal, A., Gans, J. S., & Goldfarb, A., (2017). *Spring 2017 Issue What to Expect from Artificial Intelligence* (Vol. 58). Retrieved from: http://mitsmr.com/2jZdf1Y (accessed on 21 October 2020).

Brynjolfsson, E., Mcafee, A., & Manyika, J., (2014). Will your job disappear? *New Perspectives Quarterly, 31*(2), 74–77. https://doi.org/10.1111/npqu.11457.

Crawford, K., West, S. M., & Whittaker, M., (2019). *Discriminating Systems: Gender, Race, and Power in AI*. Retrieved from: http://cdn.aiindex.org/2018/AIIndex2018AnnualReport.pdf (accessed on 21 October 2020).

Economics for the Royal Society & British Academy, (2018). *The Impact of Artificial Intelligence on Work*. Retrieved from: https://royalsociety.org/~/media/policy/projects/ai-and-work/frontier-review-the-impact-of-AI-on-work.pdf (accessed on 21 October 2020).

Gries, T., & Naudé, W., (2005). Artificial intelligence, jobs, inequality, and productivity: Does aggregate demand matter? *IZA Discussion Papers*. Retrieved from: www.iza.org (accessed on 21 October 2020).

Grosz, B. J., Altman, R., Horvitz, E., Mackworth, A., Mitchell, T., Mulligan, D., & Shah, J., (2016). *Standing Committee of the One Hundred Year Study of Artificial Intelligence*. Retrieved from: https://ai100.stanford.edu (accessed on 21 October 2020).

Jarrahi, M. H., (2018). Artificial intelligence and the future of work: Human-AI symbiosis in organizational decision-making. *Business Horizons*, *61*(4), 577–586. https://doi.org/10.1016/j.bushor.2018.03.007.

Kolb, R., & Jones, P. C., (2012). International labor organization (ILO). In: *Encyclopedia of Business Ethics and Society*. https://doi.org/10.4135/9781412956260.n438.

Kolbjornsrud, V., Amico, R., & Thomas, R., (2016). *The Promise of Artificial Intelligence: Redefining Management in the Workforce of the Future*. Accenture institute for high performance and Accenture strategy. Retrieved from: https://www.accenture.com/_acnmedia/PDF-32/AI_in_Management_Report.pdf (accessed on 21 October 2020).

OECD (2016), Skills Matter: Further Results from the Survey of Adult Skills, OECD Skills Studies, OECD Publishing, Paris. http://dx.doi.org/10.1787/9789264258051-en

Osoba, O., & Welser, W., (2017). *The Risks of Artificial Intelligence to Security and the Future of Work*. https://doi.org/10.7249/pe237.

The British Academy & The Royal Society, (2018). *The Impact of Artificial Intelligence on Work*. Retrieved from: https://www.thebritishacademy.ac.uk/sites/default/files/AI-and-work-evidence-synthesis.pdf (accessed on 21 October 2020).

Upchurch, M., (2018). Robots and AI at work: The prospects for singularity. *New Technology, Work and Employment*, *33*(3), 205–218. https://doi.org/10.1111/ntwe.12124.

Wired Insider, (n.d.). *AI and the Future of Work | Wired*. Retrieved from: https://d1ri6y1vinkzt0.cloudfront.net/media/documents/AI and the Future of Work_FIPP_VDZ.pdf (accessed on 21 October 2020).

CHAPTER 6

Understanding Consumer Responses Towards Social Media Advertising and Purchase Intention Towards Luxury Products

AMNA AHMAD[1] and BILAL MUSTAFA KHAN[2]

[1]*Research Scholar, Department of Administration, Aligarh Muslim University, Uttar Pradesh, India*

[2]*Department of Administration, Aligarh Muslim University, Uttar Pradesh, India*

ABSTRACT

The appetite for luxury goods is attributable to the inherent characteristics of luxury brands and the beneficial values (Roux, Tafani, and Vigneron, 2017) gained by having, owning, and using them (Cristini,Kauppinen-Räisänen, Barthod-Prothade, and Woodside, 2017). Luxury offerings provide outstanding quality (e.g., Choo, Moon, Kim, & Yoon, 2012) and have a more appealing appearance than non-luxury products. Luxury products are also attractive owing to features like quality materials, connoisseurship, and the core competencies of creativity, craftsmanship, and innovation that go into their making.

6.1 INTRODUCTION

As the demand for social media marketing has increased vastly through a novel promotion stand, it consents consumers to bond with one another and connect with brands. It is also observed that luxury brands' commitment to social media has also augmented because of escalation in the number of prosperous consumers who use social media. The present study surveyed the behavioral response and attitude of people to advertising and social media. Consciousness to brand was stated as one of the dynamics that influence the customer's approach to online marketing. On the other hand, it has a prominent impact on consumers' reactions to online marketing.

The approach of the consumer's towards advertising altogether held great concern for the present and previous scholars. Zanot (1984) and Elliot and Speck (1998) distinguished that the approach of consumers to online advertising has changed vibrantly during a period of time, predominantly changing from constructive to destructive. Speck and Elliot (1998) established that there are a large number of problems related to advertising, for example, hindrance in search and disruption, which causes negativity in consumer's attitudes towards advertising, to name a few. Lee (2002) and La Ferle, in research, analyzed customers' attitudes towards marketing in America as well as China and established that customers in America have a negative response to online marketing, but a majority of consumers in Chinese communities have a favorable response to online marketing. It may be stated that the attitude of consumers is inconclusive, and it majorly varies among different cultures, customs, society, and norms.

The brands, over the previous years, used social media sites-Twitter, Facebook, YouTube, and Pinterest as channels to endorse themselves and to connect to the consumers. Contemporary prognoses anticipate that the worldwide expenses on online marketing increased to $8.09 billion in the year 2012, which is 10.2% of worldwide virtual advertisement expenditure (Lee, Lee, and Choi, 2012). In spite of the developing group of studies on investigative online marketing (e.g., Hadija, Taylor, Lewin, and Strutton, 2011; Hair et al., 2012), there are very few experimental market research that studies the efficacy of online marketing if we talk in the horizon of luxurious products. Therefore, the question upturned here: in what ways online marketing activities influence consumers' buying preference to luxury brands?

As demarcated by Interbrand (2008), a luxurious brand "demonstrates a sense of snootiness, uniqueness, extravagance, and exclusivity." Over the last few years, luxury products did gradually embrace social media marketing activities (Phan, 2011). However, few are of the opinion that large accessibility of social media is conflicting in the areas of the exclusiveness of luxurious brands, but connoisseurs from the trade are of the opinion that social media and its marketing tactics stimulate customer's commitment to luxurious products (Ortved, 2011; Gers, 2009).

6.2 LITERATURE REVIEW

From contemporary learning, Kamal and Chu (2011) inspected social media and its impact, brand cognizance, and buying intent to luxury brands on consumers. They also studied the responses that people have to marketing tactics of social media. Kamal and Chu (2011) established that online media practice influences people in an assertive way. Moreover, results obtained from their study determined that consumers who are more inclined towards brand awareness have promising opinions and attitudes to online advertising, as against the people who are less inclined towards brand cognizance. Observing the trend in the speedy acceptance as well as the admiration of luxury brands promotion in online media, there is a requisite for empirical research to be conducted to examine consumers' attitudes, beliefs also behavioral reactions towards social media promotion. The research on this subject is exclusive since it takes into account-whether brand consciousness is just an alternative to advertising credence and the way it associates to consumers' approaches towards social media marketing. It is well established that brand awareness holds a special element to be observed, because it has an enormous effect towards purchase choices towards luxurious brands.

6.3 SOCIAL MEDIA AND LUXURY BRANDS

As per statistics in 2011-the amount of prosperous social media users increased to 57.1 million (eMarketer, 2007) from 43.7 million in the USA in the year 2006. The rise in the percentage of affluent social media consumers impacts the demand of luxury brands and sales of elite products (Okonkwo, 2009). As per the statistics in 2011 by Unity, 78% of prosperous buyers are users of social media; also, 50% of the consumers use online media to pursuit info about luxury brands. In 2016, the international market for luxurious products had reached €318 billion (Choi, Seo, Wagner, and Yoon, 2018). As it is noted that 75% of luxury consumers practice social media, hence, the majority of luxury brands like Chanel, Burberry, etc., are now making their foray into social media marketing to connect with their clients because it provides a platform to intensify brand consciousness, penchant, and behavioral intents (Godey et al., 2016; Seo and Shin, 2017).

In the current scenario, luxury brands like Chanel, Burberry, to name a few, use social media to construct associations to the buyers also to augment sales and brand experience (Kim and Ko, 2010; Corcoran and Feugere, 2009;

Phan, Thomas, and Heine, 2011). To take an example, Burberry was branded as a social media initial adopter, as it had around 800,000 admirers on Twitter (social media site) and above 11.60 million "followers" on social sites like Facebook (Messieh, 2012). Other luxury brands, for example, Chanel, Gucci, Louis Vuitton, to name a few, are also getting involved in marketing strategies of online advertising. The profile pages of high-end luxurious products via social media platforms offer consumers insight into the history of the luxury brand also emphasizes the refinement, distinctiveness, and exclusivity of the brand. The profile pages of these exclusive brands permit luxury dealers to get associated with their fine clients on an individual level also deliver genuine-time info regarding brands, something which is of particular not feasible through company sites. Profile pages of the brands also inspire word-of-mouth dialogues amongst customers (Peters and Thomas, 2011), which offers an opportunity for luxury brand dealers to clout luxury brand sponsors. Ko and Kim (2011), of late, brought into being the optimistic impact of brands' online advertising on client's associations as well as on the purchase intention of these elite, prosperous, high-class, and exclusive clients.

6.4 CONSUMER'S BELIEFS TO STRATEGIES OF ONLINE ADVERTISEMENT

Social media and its marketing tactics give a fine chance to dealers-a platform-to make available an assortment of information which can be used by the users, at their own pace, in their own space, in their own time, and in their own comfort. If we take an instance on the previous researches of social media promotion (Ducoffe, 1996), it is presumed that online marketing will offer consumers the latest information because of the perception of the unique, exclusive, and elite consumer's towards online advertising informativeness would become an essential predecessor. Hence, informativeness is important in forming a constructive attitude, belief, and approach of consumers to social media promotion.

Next, materialism is taken into account to explore its effect into shaping consumer's beliefs. Materialism basically reconnoiters consumer's beliefs that put emphasis on how consumption provides a platform for satisfaction and fulfillment to users (Pollay and Mittal, 1993). Social media operators are often viable to recurrent consumption-linked posts; this is a major reason that induces a greater level of covetousness, avarice, acquisitiveness, and materialism amongst consumers. It is well noted that exposure through advertisement tempts materialism and covetousness (Chia and Jiang,

2009); materialism may imply as a core belief dynamic which deleteriously associates with online promotion attitudes, beliefs also norms.

The credibility of online advertising used by brands is of utmost importance to clients. The falsity belief (Pollay and Mittal, 1993) is also an imperative element for users' attitudes to social media marketing. To make online advertising more trustworthy and credible, marketers should first check the authenticity of the advertised message. The current study, hereby, incorporates the falsity dynamic to study the ideologies of consumers to the ill influences of online marketing on social ethics. It would build the consumer's attitude to online marketing.

Last of all, value corruption (Pollay and Mittal, 1993) also plays a key role in reconnoitering consumers' thoughts on social media advertising's aptitude also the way it contours the values of the buyer's. It sometimes leads to endorsing behavior and attitudes that might not be acceptable or conventional in society. It is also interesting to note that Wolin, Korgaonkar, and Lund (2002) were of the opinion that advertisements through social media might perhaps endorse or support messages that might not be as per the decorum, dignity, propriety of the societal standards, norms, values, ideals, and beliefs. If taken from the perspective and viewpoint of the user's, value corruption dogmas perhaps are allied adversely to online publicizing or social media marketing, that possibly be deliberated further "under the radar" as compared to explicit, overt, and open nature of media, for instance, mass media platform. Altogether, these beliefs could well be said to be concomitant with attitudes, which in that way, all together, has an impact on people's beliefs, outlook, opinion, and attitude to social media and its marketing strategies.

6.5 BRAND AWARENESS OR CONSCIOUSNESS

Brand consciousness can be noted as an imperative area in studying people's attitudes and their psychology (Shim and Gehrt, 1996; McLeod and Nelson, 2005). Consciousness towards a brand can be described as having a high regard for specific brands and making efforts to acquire knowledge about the exquisite brands (Nelson, Shah, Keum, and Devanathan, 2004). This behavior of users effect their attitude to social media marketing as people who display an assertive attitude to a brand also displays an assertive attitude to online advertising. Brand consciousness hence is considered an important factor in determining consumer's attitudes to online advertising.

6.6 PEOPLE'S ATTITUDE TO ONLINE MARKETING OF LUXURY BRANDS

Studies that pertain to consumer behavior suggest that consumer response follows a pattern towards social media marketing (Stiener and Lavidge, 1961). Hence, the behavior of users might be ordered in affective, cognitive area, also behavioral patterns. In the beginning, consumers enter the phase of cognition; in this, users gain awareness about the luxury product. Then comes the affective phase, in this consumer's form a preference, fondness, also persuasion towards a luxury brand. Finally, it leads to the purchase of the brand. Hence, the beliefs of people to social media and its advertising play a fundamental role in determining user's beliefs on social media (Ducoffe, 1996). Thereby, beliefs about advertising are an important element to define users' attitudes toward online advertising.

6.7 BELIEFS, BEHAVIORAL RESPONSES, ALSO PURCHASE BEHAVIOR TO LUXURY BRANDS

As stated by Mehta (2000) that attitudes towards advertising are mostly related in a constructive manner to brand reminiscence, purchasing intention, and aim. To lay emphasis on social media advertising, Bruner and Kumar (2000) established that people who have a negative approach to social media marketing notched a lower assessment in the dynamics of purchase intent as compared to those who have an affirmative response to social media marketing.

As stated by Korgaonka and Wolin (2002), the beliefs of online marketing may possibly be associated with the behavioral interests of the consumers as well as negative beliefs on social media. It might be put forth that the consumers who have a positive attitude to luxury brand pages in social media would involve vigorously in the advertisement and pursue info regarding brands that are being promoted or publicized. It directly or indirectly has an impact on snowballing the purchasing intention of luxury brands. For instance, Burberry–the brand that is most "adored" luxury brand on social media-having fans of hardcore loyalty–involved motivated and genuine clients in the social campaigns conducted by them and witnessed an increase of 21% in earnings (Messieh, 2012). Consequently, an experimental study shows that the behavioral responses and reactions towards online advertising might be associated towards consumers' attitudes to procure luxurious brands.

The objective of the paper is to deliver advertising inferences in relation to the marketing strategies of social media and ways to involve people, also how online marketing tactics impact buying behavior of luxurious products. It also studies the response of users to social media and its marketing. To say in precise is the fact that users have a progressive attitude and opinion to social media; hence, the elite brands like Chanel, Hermes, etc., should use online marketing as a form of promotion. In addition, it can be proposed that to increase their credibility, and luxurious products ought to consider strategies to increase the candor of the company and the ways to use online promotion as a marketing custom. Certain strategies to suggest that may help the luxury brands to accomplish inordinate perceived integrity with the users: updating the content of social media promotion that would provide assuredness that the facts provided online are contemporary, also forming connections with superior brands and companies. To say in a precise manner, luxury brands that inspire to provoke progressive behaviors from users and their loyal consumers should make sure that the social media public relations are providing up-to-date, reliable, and accurate information. Advertisements should also highlight decent societal values, beliefs, and norms.

As it may be inferred that social media and its marketing strategies are a boon to luxury brands as it is a means to endorse and inform consumers about their brands and products. Furthermore, researchers in the past recommend that brand-conscious entities are more concerned with how they are perceived by others and are inclined towards fashion and style(Bush, Bloch, and Dawson, 1989). The findings of the paper also suggest that by networking with social media marketing, consumers may connect to their online followers (Litt, 2012) associate with the brands and may share the messages amongst their peers-friends, family, and co-workers. The markers permit brand-loyal consumers to involve themselves in self-management also provide an excellent platform to find out more about the brands.

6.8 CONCEPTUAL MODEL

A conceptual model has been put forth based on the study. It takes into account the key variables (see Annexure 1; Figure A1. Conceptual model).

Social media are computer-mediated expertise which helps in the conception and distribution of information. It also is beneficial in professional aspects and career orientations. Web/Social/Virtual media may be put forth as virtual presentations, mass media, and platforms that help in collaborations, associations, and assists in sharing of matter (Richter and Koch, 2007). Social media have a number of categories, some of which are social blogs, wikis, microblogs, pictures, weblogs, video, rating, podcasts, and social bookmarking. The advancement in technology is a great advantage for luxury brands as it assists in enticing consumers to interrelate with these brands and form a kind of network with them. The engagement of luxury brands in activities like writing a blog, tweeting, and networking made these luxury brands to keep up-to-date with the contemporary style. The luxury brand is often labeled as the phantasmagorias in the minds of customers that lead to forming a connotation about its elite pricing, quality, aesthetics, paucity, extraordinariness, uniqueness, exclusivity, and non-functional associations.

Purchase intention may be stated as an amalgamation of customers' interest that leads to the purchase of the brand. Consumers buying interest may be put onwards as a behavioral construct that serves as a benchmark for determining clients' future contributions towards the brand; however, customer equity is a behavioral construct that takes into account the actual purchasing behavior of the customers and their loyalty towards the luxury brand. This also plays a noteworthy role in keeping a tab on consumers' future buying behavior (Park, Ko, and Kim, 2010).

6.9 METHODOLOGY

India is a developing nation, and the consumers in India are involved in the purchase of luxury brands (Bhardawaj et al., 2005). The research focuses on consumers who follow a decent lifestyle and have an inclination towards luxury goods.

The questionnaire was administered in six cities-Bangalore, Delhi, Noida, Gurgaon, Faridabad, and Ghaziabad. Three shopping malls were selected from the cities mentioned above, and the shoppers were entreated to fill up the questionnaire while they were shopping in the mall.

6.10 INSTRUMENT

The survey method was used to explore the consumer's response through the five variables (product information belief, materialism, value corruption, falsity, brand consciousness/awareness) and how it influences the purchase of the luxury brand. To fulfill the mentioned objective, a questionnaire was made and tested on 60 respondents (pilot survey) in the New Delhi-NCR region to measure the accuracy, reliability, and validity of the sample.

The items assimilated in the survey focused on the areas of online marketing, consumer attitude, and also their buying behavior to the luxury brand. Furthermore, items covering the aspects of demographics were also incorporated. The responses of consumers were noted on a Likert scale measuring 5 points.

6.11 DATA ANALYSIS

Around 600 questionnaires were circulated, but the responses considered valid for the study were 472. Initially, the reliability and validity of the data were verified. Then, the hypothesis was tested. The root mean square residual (RMR) comparative fit index (CFI), the goodness of fit (GFI), also root mean square error of approximation (RMSEA), and adjusted goodness of fit (AGFI) were used on structural as well as measurement model in accordance with the model fit.

6.12 SAMPLE CHARACTERISTICS

Excel and SPSS23 were used to measure the characteristics of the sample. The resultant tables and the conclusions give a gist of the characteristics of the participants. Overall the number of respondents gives a representation of both the gender-be it female or male, majorly all the age clusters were qualified to be educated.

A detailed analysis of the characteristics of the sample of respondents on gender, qualification, age group, employment status, and family income are given in Annexure 2 (Table A1. Summary of sample characteristics on the basis of gender).

As seen from the results mentioned above, around 472 participants are youth and in the middle age group with a decent monthly family income. Both the genders-male and female represent an equal proportion, i.e., at 54.9%, 45.1%, and 46% of the participants are in the age group 25–34 years that's a characteristic of the sample size; also it signifies that a major proportion of the population is young.

Overall, it can be stated that the population studied is well demonstrative and very appropriate for the luxury brand study anticipated.

6.13 NORMALITY (SKEWNESS AND KURTOSIS)

Statistical tests used to evaluate normality are kurtosis and skewness. Kurtosis and skewness should be within the +2 to –2 range (Hair, 2006). Descriptive statistics that include skewness, as well as kurtosis, were performed using SPSS 23 on the eight variables. The values that have been measured for kurtosis and skewness of all the eight variables lie between +3 to −3 range concluding-sample has a normal distribution.

The scale statistics obtained using SPSS 23 is illustrated in Annexure 3; Table A6. Construct statistics-Skewness and Kurtosis.

6.14 RELIABILITY

"Reliability can be stated as the amount as to experimental construct measures 'actual' value and does not contain any error. This is contradictory to the error in measurement. On repeated trials, results with more reliability would exhibit a larger consistent value as compared to the measures with lesser reliability" (Hair, 2014, p. 8).

"The measurement that assesses those variables is measuring the quantity that is expected to be measured. Also, the reliability takes into account the consistency of the data" (Hair et al., 2014, p. 2).

"The reliability coefficient that is also called Cronbach's alpha measures the consistency of the data and is the most popular scale to be used. A lower limit of 0.5 is considered acceptable while calculating the exploratory factor analysis" (Hair et al., 2006, p. 161).

The reliability coefficient was analyzed for the factors (42 items) with the help of SPSS 23.

The result is illustrated in Annexure 4; Table A7. Cronbach's alpha of constructs.

The variables that showed the highest reliability were Social media advertising attitude (0.883), product information belief (0.831), purchase intention towards luxury products (0.812), and falsity (0.801). The other variables were all above 0.7, except social media advertising behavior. All the variables were accepted as they were above 0.5.

6.15 FACTOR ANALYSIS

Degree of sample acceptability-Kaiser-Mayer-Olkin (KMO) also a test of sphericity-Bartlett's were applied to determine the suitability of statistics for the analysis of factors. The PCA method (for factor analysis) or The Principal Component Analysis by means of (varimax) orthogonal-rotation was used to maintain constructs, and factor loading was established to be above 0.4 besides Eigen standards over 1. An item is removed, conditioned it doesn't load over 0.40 on its variable (Hair, 1998).

6.15.1 MOTIVATION MEASURE FOR ANALYSIS OF FACTORS

The KMO, as well as the test of sphericity by Bartlett's, were used for motivation extent-presented in Table A8; Annexure 5.

The KMO value of sampling acceptability should be more than 0.4 for adequate study. From the values in the table, it is obvious that the KMO degree of sampling acceptability value of 0.936 is more than acceptable, and Bartlett's test for sphericity is moreover acceptable as the related probability is below 0.05(0.000).

6.15.2 CONSTRUCT VALIDITY

Hair (2006) explains the construct validity: "set of items measure the latent variables that were intended to be measured."

In order to calculate the construct validity, the subsequent modules could be evaluated that may attain the validity of variable; these are the following recommendations:

1. **Factor Loading:** The rule that's followed is that the consistent loadings ought to be 0.40 or greater.
2. **The Average Variance Extracted:** A value of AVE equal to 0.5 or higher is considered appropriate representing adequate convergence.
3. **Construct Reliability:** Reliability value greater than 0.7 is considered adequate.

On the basis of the above-mentioned guidelines, a model was analyzed with the help of SPSS 23 to analyze the factor loadings of the constructs and their construct validity. The results hence obtained are presented in Table A9; Annexure 6.

6.16 MEASUREMENT MODEL FIT INDICES

The way to test measurement theory is by comparing the theoretic measurement model contrary to realism, as signified by the sample. This is completed by reviewing basic GOF (goodness-of-fit) to measure the fit.

6.17 STRUCTURAL EQUATION MODEL

As can be seen from Table 7 that out of seven, five hypotheses strongly support product information belief, materialism, falsity, behavior towards online marketing. Therefore, it might be put forth that the above-mentioned constraints ought to have a pronounced influence on the buying behavior and not value corruption and brand consciousness (see Annexure 8; Table A11).

- *H1: Social media advertising attitude has a positive impact on product information belief.*
  H1 is accepted.
- *H2: Social media advertising attitude has a positive impact on materialism.*
  H2 is accepted.
- *H3: Social media advertising attitude has a positive impact on falsity.*
  H3 is accepted.
- *H4: Social media advertising attitude has a positive impact on value corruption.*
  H4 is not accepted.
- *H5: Social media advertising attitude has a positive impact on brand consciousness.*
  H5 is not accepted.

6.18 HYPOTHESIS TESTING AND ANALYSIS

We have collected data from different cities and conducted tests through Spss and amos. On the basis of which we have analyzed the hypothesis (See Annexure 9; Table A12).

6.19 CONCLUSION

With the dawn of online media, communication has become a lot easier for consumers. In a study conducted by Mayzlin and Godes (2004), it was studied that online marketing sites are highly lucrative and a great platform to assess information from customers. Hence, social media are very advantageous to luxury brands as it helps them to maintain an association with their clients. Bruyn and Lilien (2008), in research conducted by them, studied the impact of consumer communications and how it aids through the making of decisions in the framework of viral marketing.

The tactics and strategies used by brands to form an association with their clients have evolved significantly over the last decade because of the introduction of online networking sites–Twitter, Pinterest, Instagram, and many more. Social media can be defined as the contents that are uploaded by customers or marketers–video, image, slide show or blog, etc. The online media provides a platform for customers to interrelate with each other, exchange information, and ideas in a cybernetic zone (Wikipedia). An important aspect of online media is-it provides two-mode communication; therefore, it is also called consumer-generated media (Rob, 2009).

The present study investigates the influences of online marketing on buyers' attitudes in the context of the luxurious brand, also how it affects their purchase intention.

The paper studies the influence of different factors (value corruption, product information, belief, materialism, and falsity) on the attitude of users to luxury products and their buying behavior. The influence of these factors was compared across different variables.

ANNEXURE 1

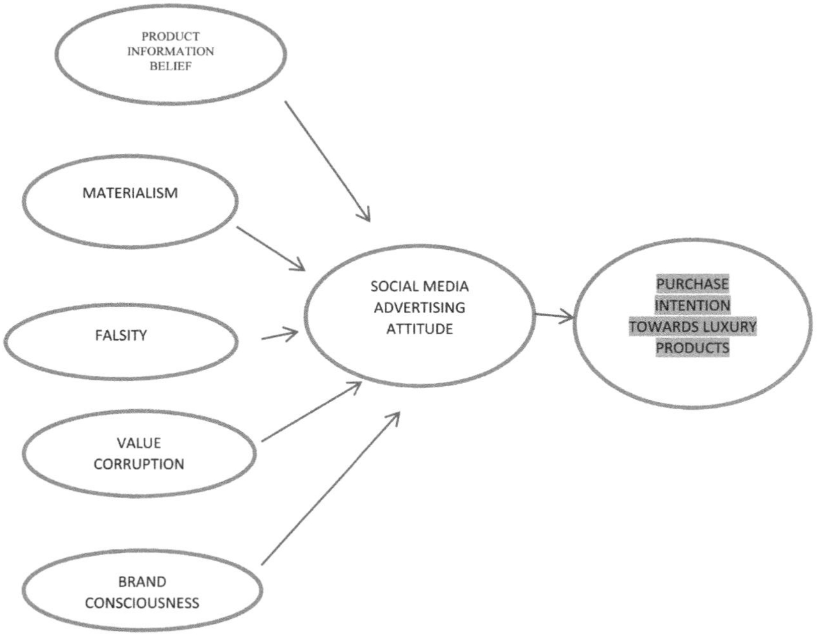

FIGURE A1 Conceptual model.

ANNEXURE 2

> **Gender:** Of the total of 472 respondents 259 are males (54.9%) and 213 are females (45.1%) (n=472). Both the females and the male population are shown in a fair manner in the study, as is represented in Table A1.

TABLE A1 Summary of Sample Characteristics on the Basis of Gender

		Gender			
		Frequency	Percent	Valid Percent	Cumulative Percent
Valid	1	259	54.9	54.9	54.9
	2	213	45.1	45.1	100.0
	Total	472	100.0	100.0	

1 = Male, 2 = Female.
Source: The author.

> **Qualification:**

TABLE A2 Summary of Sample Characteristics-Qualification

		Qual.			
		Frequency	Percent	Valid Percent	Cumulative Percent
Valid	1.0	82	17.4	17.4	17.4
	2.0	205	43.4	43.4	60.8
	3.0	95	20.1	20.1	80.9
	4.0	78	16.5	16.5	97.5
	5.0	12	2.5	2.5	100.0
	Total	472	100.0	100.0	

Source: The author.

> **Employment Status:**

TABLE A3 Gist of Sample Wise Data-Employment

		Emplstat.			
		Frequency	Percent	Valid Percent	Cumulative Percent
Valid	1.0	179	37.9	37.9	37.9
	2.0	17	3.6	3.6	41.5
	3.0	39	8.3	8.3	49.8
	4.0	20	4.2	4.2	54.0
	5.0	194	41.1	41.1	95.1
	6.0	1	0.2	0.2	95.3

TABLE A3 *(Continued)*

		Emplstat.			
		Frequency	Percent	Valid Percent	Cumulative Percent
	7.0	6	1.3	1.3	96.6
	8.0	13	2.8	2.8	99.4
	9.0	3	0.6	0.6	100.0
	Total	**472**	**100.0**	**100.0**	

Source: The author.

> **Age Group:** Majority of respondents exceeded the age of bracket-18 years. A total of 162 respondents (34.3%) fall in the age bracket of 18–24 years of age. The frequency of 25–34 years of age group falls to 217 respondents (46.0%), 35–44 age group to 84 (17.8%), 35–44 to 84 (17.8%), 45–54 years of age group to 4 (0.8%), 55–64 years of age group to 1 (0.2), 65–74 years of age group to 2 (0.2%) and finally above 75 falls to 1 (0.2%).
>
> As evident from the data, the maximum falls in the category of 25–34 years of age, representing that the younger population is brand conscious as well as fashion conscious.

TABLE A4 Summary of Sample Characteristics-Age group

		Frequency	Percent	Valid Percent	Cumulative Percent
Valid	1.0	1	0.2	0.2	0.2
	2.0	162	34.3	34.3	34.5
	3.0	217	46.0	46.0	80.5
	4.0	84	17.8	17.8	98.3
	5.0	4	0.8	0.8	99.2
	6.0	1	0.2	0.2	99.4
	7.0	2	0.4	0.4	99.8
	8.0	1	0.2	0.2	100.0
	Total	**472**	**100.0**	**100.0**	

1 = Below 18 years, 2 = 18–24 years, 3 = 25–34 years, 4 = 35–44 years, 5 = 45–54 years, 6 = 55–64 years, 7 = 65–74 years, 8 = 75 years or older.
Source: The author.

> **Family Income (Monthly):** The family income is spread across the slabs with Rs.50,000 to Rs.1,00,000 having the maximum share at 30.5% (n=144). The salaries represent the ongoing status.

Table A5 gives the following analysis:

TABLE A5 Summary of Sample Characteristics-Family Income (Monthly)

		Frequency	Percent	Valid Percent	Cumulative Percent
Valid	1.0	144	30.5	30.5	30.5
	2.0	105	22.2	22.2	52.8
	3.0	43	9.1	9.1	61.9
	4.0	42	8.9	8.9	70.8
	5.0	138	29.2	29.2	100.0
	Total	472	100.0	100.0	

1 = Less than Rs.50,000; 2 = Rs.50,000–Rs.1,00,000; 3 = Rs.1,00,000–Rs.1,50,000; 4 = Rs.1,50,000–Rs.2,00,000; 5 = More than Rs.2,00,000.

Source: The author.

ANNEXURE 3

TABLE A6

	PIB	M	F	VC	BC	SMAA	PILP
N	472	472	472	472	472	472	472
Valid	472	472	472	472	472	472	472
Missing	0	0	0	0	0	0	0
Std. deviation	0.67382	0.76601	0.64736	0.79717	0.69287	0.68646	0.68856
Skewness	−1.098	−0.105	−0.386	−0.384	−0.365	−0.770	−0.710
Std. error of skewness	0.112	0.112	0.112	0.112	0.112	0.112	0.112
Kurtosis	1.335	−0.235	0.063	−0.067	0.177	0.646	0.682
Std. error of kurtosis	0.224	0.224	0.224	0.224	0.224	0.224	0.224

ANNEXURE 4

TABLE A7

No.	Construct	Cronbach's Alpha	Number of Items
1.	Product Information Belief	0.831	6
2.	Purchase Intention Towards Luxury Products	0.812	5
3.	Social Media Advertising Attitude	0.883	7
4.	Materialism	0.762	4
5.	Falsity	0.801	6
6.	Value Corruption	0.799	4
7.	Brand Consciousness	0.728	5

ANNEXURE 5

TABLE A8

KMO and Bartlett's Test		
Kaiser-Meyer-Olkin Measure of Sampling Adequacy.		.936
Bartlett's Test of Sphericity	Approx. Chi-Square	8254.173
	df	666
	Sig.	.000

ANNEXURE 6

TABLE A9

No.	Latent Variable Measured	Item Description	Std. Loading	AVE	CR
Product information belief				0.459	0.835
1.	PIB1	Social media advertising is a very valuable source of information about sales.	0.707		
2.	PIB2	Social media advertising tells me which brands have the features I am looking for.	0.704		
3.	PIB3	Social media advertising helps me keep up to date about products available in the marketplace.	0.669		
4.	PIB4	I think the information obtained from the social networking sites ad would be helpful.	0.714		
5.	PIB5	I would like to pass out information on brands, products, or services from social media sites to my friends.	0.656		
6.	PIB6	It is fun to collect information on brands or fashion items through social media.	0.553		
Purchase intention towards luxury products				0.483	0.821
1.	PILP1	I am willing to recommend others to buy a luxury product/brand.	0.697		
2.	PILP2	I intend to purchase a luxury product/brand in the future.	0.826		
3.	PILP3	I will buy a product/service advertised on social media.	0.592		
4.	PILP4	I would consider buying a luxury product.	0.672		
5.	PILP5	If I am in need, I will buy a luxury product.	0.570		

TABLE A9 *(Continued)*

No.	Latent Variable Measured	Item Description	Std. Loading	AVE	CR
	Social media advertising attitude			0.524	0.885
1.	SMAA1	I consider social media advertising very important.	0.714		
2.	SMAA2	Advertisements through social media are more interesting than traditional advertising.	0.629		
3.	SMAA3	Social media advertising is more interactive than traditional advertising.	0.655		
4.	SMAA4	Overall, I consider social advertising a good thing.	0.674		
5.	SMAA5	Overall, I like social media advertising.	0.624		
6.	SMAA6	I consider social media advertising very essential.	0.624		
7.	SMAA7	I would describe my overall attitude toward social media advertising very favorably.	0.506		
	Materialism			0.460	0.764
1.	M1	Social media advertising increases dissatisfaction among consumers by showing products which some consumers can't afford.	0.705		
2.	M2	Social media advertising is making us a materialistic society-interested in buying and owning things.	0.776		
3.	M3	Social media advertising makes people buy unaffordable products just to show off.	0.761		
4.	M4	Social media advertising leads to luxury brand culture.	0.468		
	Falsity			0.412	0.805
1.	F1	One can put more trust in products advertised on social media than in those not advertised on social media.	0.560		
2.	F2	Certain products play an important role in my life.	0.756		
3.	F3	Social media advertisements reassure me that I am doing the right thing in using luxury products.	0.607		
4.	F4	Social media advertising helps the consumer buy the best brand for the price.	0.681		

TABLE A9 *(Continued)*

No.	Latent Variable Measured	Item Description	Std. Loading	AVE	CR
5.	F5	I trust the information shared with me by people I know through social media channels.	0.543		
6.	F6	E-commerce companies who are well known in social media are credible.	0.485		
Value corruption				0.518	0.808
1.	VC1	Social media advertising takes undue advantage of people.	0.739		
2.	VC2	Social media advertising leads people to make unreasonable purchase demands.	0.787		
3.	VC3	Social media advertising sometimes makes people live in a world of fantasy.	0.671		
4.	VC4	There is a lot of inappropriate stuff on social media nowadays.	0.671		
Brand consciousness				0.356	0.729
1.	BC1	I prefer buying the best-selling brands.	0.445		
2.	BC2	The more expensive brands are usually my choices.	0.451		
3.	BC3	Nice department and specialty stores offer me the best products.	0.672		
4.	BC4	I am willing to pay a higher price for the brand I prefer to buy.	0.440		
5.	BC5	The price of my favorite brand would have to increase quite a bit before I would switch to another brand.	0.496		

ANNEXURE 7

TABLE A10

Fit Indices	Recommended Value	Source	Observed Value
CMIN	<3.0	Kline (2004)	2.5
AGFI	>=9.0	Bentler and Bonett (1980)	0.799
CFI	>0.90	Bentler and Bonett (1980)	0.871
RMSEA	<0.07	Steiger (2007)	0.056
PCLOSE	>0.05	Hair et al. (2010)	0.000
PNFI	No cut off value	Hair et al. (2008)	0.738
GFI	>=0.90	Joreskog and Sorenbem (2002)	0.824

ANNEXURE 8

TABLE A11

			Estimate	S.E.	C.R.	P	Result
SMAA	<---	PIB	0.559	0.026	21.334	***	Supported
SMAA	<---	M	−0.049	0.023	−2.106	0.035	Supported
SMAA	<---	F	0.375	0.027	13.746	***	Supported
SMAA	<---	VC	−0.008	0.022	−0.373	0.709	Not Supported
SMAA	<---	BC	0.029	0.025	1.138	0.255	Not Supported

ANNEXURE 9

TABLE A12

H1: Gender has a positive impact on product information belief

	F	Sig.	t	df	Sig. (2-tailed)
PIB Equal variances assumed	13.871	0.000	−3.754	470	0.000
Equal variances not assumed			−3.848	468.24	0.000

As the significance is less than 0.05; hence, the null hypothesis is accepted.

H2: Gender has a positive impact on materialism.

	F	Sig.	t	df	Sig. (2-tailed)
M Equal variances assumed	1.454	0.229	−1.467	470	0.143
Equal variances not assumed			−1.478	463.51	0.140

As the significance is more than 0.05; hence, the null hypothesis is not accepted.

H3: Gender has a positive impact on falsity belief.

	F	Sig.	t	df	Sig. (2-tailed)
F Equal variances assumed	4.949	0.027	0.259	468	0.796
Equal variances not assumed			0.262	466.43	0.793

As the significance is greater than 0.05; hence, the null hypothesis is not accepted.

H4: Gender has a positive impact on value corruption.

	F	Sig.	t	df	Sig. (2-tailed)
VC Equal variances assumed	0.587	0.444	−1.322	468	0.187
Equal variances not assumed			−1.327	455.10	0.185

As the significance is greater than 0.05; hence, the null hypothesis is not accepted.

H5: Gender has a positive impact on brand consciousness.

	F	Sig.	t	df	Sig. (2-tailed)
BC Equal variances assumed	2.948	0.087	1.857	468	0.064
Equal variances not assumed			1.880	464.87	0.061

As the significance is greater than 0.05; hence, the null hypothesis is not accepted.

H6: Gender has a positive influence on social media advertising attitude.

	F	Sig.	t	df	Sig. (2-tailed)
SMAA Equal variance assumed	7.390	0.007	−2.147	468	0.032
Equal variances not assumed			−2.184	467.41	0.029

As significance is less than 0.05; hence, the null hypothesis is accepted.

H7: Gender has a positive impact on purchase intention towards the luxury product.

	F	Sig.	t	df	Sig. (2-tailed)
PILP Equal variances assumed	0.466	0.495	−0.420	468	0.674
Equal variances not assumed					0.672

As the significance is greater than 0.05; hence, the null hypothesis is not accepted.

KEYWORDS

- **adjusted goodness of fit**
- **comparative fit index**
- **goodness of fit**
- **Kaiser-Mayer-Olkin**
- **root mean square error of approximation**
- **root mean square residual**

REFERENCES

Bush, R. F., Bloch, P. H., & Dawson, S., (1989). Remedies for product counterfeiting. *Business Horizons, 32*, 59–65.

Chevalier, J. A., & Mayzlin, D., (2006). The effect of word of mouth on sales: Online book reviews. *Journal of Marketing Research, 43*(3), 345–354.

Choi, Y. K., Seo, Y., Wagner, U., & Yoon, S., (2018). Matching luxury brand appeals with attitude functions on social media across cultures. *Journal of Business Research*.

Chu, S. C., & Kamal, S., (2011). An investigation of social media usage, brand consciousness, and purchase intention towards luxury products among millennials. In: Okazaki, S., (ed.), *Advances in Advertising Research: Breaking New Ground in Theory and Practice* (Vol. 2, pp. 179–190). Wiesbaden: Gabler Verlag.

Corcoran, C. T., & Stephane, F., (2009). Brands aim to adapt to social media world. *WWD: Women's Wear Daily, 198*, 20–21.

De Bruyn, A., & Lilien, G. L., (2008). A multi-stage model of word-of-mouth influence through viral marketing. *International Journal of Research in Marketing, 25*(3), 151–163.

Ducoffe, R. H., (1996). Advertising value and advertising on the web. *Journal of Advertising Research, 36*, 21–35.

Gers, D., (2009). *Social Climbing: Luxury Fashion Brands Must Embrace Social Media*. Forbes. Retrieved from: http://www.forbes.com/2009/10/14/social-media-luxury-brands-cmo-networkgers.html (accessed on 21 October 2020).

Godes, D., & Dina, M., (2004). Firm-created word-of-mouth communication: A field-based quasi-experiment. *HBS Marketing Research Paper,* 04–03.

Godey, B., Manthiou, A., Pederzoli, D., Rokka, J., Aiello, G., Donvito, R., & Singh, R., (2016). Social media marketing efforts of luxury brands: Influence on brand equity and consumer behavior. *Journal of Business Research, 69*(12), 5833–5841.

Hadija, Z., Barnes, S. B., & Hair, N., (2012). Why we ignore social networking advertising. *Qualitative Market Research: An International Journal, 15*, 19–32.

Interbrand, (2008). *The Leading Luxury Brands-2008*. Retrieved from: http://www.interbrand.com (accessed on 21 October 2020).

Jiang, R., & Chia, S. C., (2009). The direct and indirect effects of advertising on materialism of college students in China. *Asian Journal of Communication, 19*, 319–336.

Keum, H., Devanathan, N., Deshpande, S., Nelson, M. R., & Shah, D. V., (2004). The citizen consumer: Media effects at the intersection of consumer and civic culture. *Political Communication, 21*, 369–392.

Kim, A. J., & Ko, E., (2010). Impacts of luxury fashion brand's social media marketing on customer relationship and purchase intention. *Journal of Global Fashion Marketing, 1*, 164–171.

Lavidge, R. J., & Steiner, G. A., (1961). A model for predictive measurements of advertising effectiveness. *Journal of Marketing, 25*, 59–62.

Litt, E., (2012). Knock, knock, who's there? The imagined audience. *Journal of Broadcasting and Electronic Media, 56*, 330–345.

Mehta, A., (2000). Advertising attitudes and advertising effectiveness. *Journal of Advertising Research, 40*, 67–72.

Messieh, N., (2012). *How Luxury Brands are Using Social Media*. Retrieved from: http://thenextweb.com/socialmedia/2012/03/20/luxury-brands-using-social/ (accessed on 21 October 2020).

Nelson, M. R., & Devanathan, N., (2006). Brand placements bollywood style. *Journal of Consumer Behavior, 5*, 211–221.

Nelson, M. R., & McLeod, L., (2005). Adolescent brand consciousness and product placements: Awareness, liking and perceived effects on self and others. *International Journal of Consumer Studies, 29*, 515–528.

Okonkwo, U., (2009). Sustaining the luxury brand on the internet. *Journal of Brand Management, 16*, 302–310.

Phan, M., (2011). Do social media enhance consumer's perception and purchase intentions of luxury fashion brands? *The Journal for Decision Makers, 36*, 81–84.

Phan, M., Thomas, R., & Heine, K., (2011). Social media and luxury brand management: The case of Burberry. *Journal of Global Fashion Marketing, 2*, 213–222.

Pollay, R. W., & Mittal, B., (1993). Here's the beef: Factors, determinants, and segments in consumer criticism of advertising. *Journal of Marketing, 57*, 99–114.

Quigley, M., Conley, K., Gerkey, B., Faust, J., Foote, T., Leibs, J., & Ng, A. Y., (2009). ROS: An open-source robot operating system. In: *ICRA Workshop on Open Source Software* (Vol. 3, No. 3.2, p. 5).

Shim, S., & Gehrt, K. C., (1996). Hispanic and Native American adolescents: An exploratory study of their approach to shopping. *Journal of Retailing, 72*, 307–324.

Stevenson, J. S., Bruner, G., & Kumar, A., (2000). Web page background and viewer attitudes. *Journal of Advertising Research, 20*, 29–34.

Taylor, D. G., Lewin, J. E., & Strutton, D., (2011). Friends, fans, and followers: Do ads work on social networks? How gender and age shape receptivity. *Journal of Advertising Research, 51*, 258–275.

Thomas, J. B., & Peters, C. O., (2011). Which dress do you like? Exploring brides' online communities. *Journal of Global Fashion Marketing, 2*, 148–160.

Wolin, L. D., Korgaonkar, P., & Lund, D., (2002). Beliefs, attitudes, and behavior towards web advertising. *International Journal of Advertising, 21*, 87–113.

CHAPTER 7

Multichannel Banking and Customer Experience: A Literature Review of Channels as a Moderator

NIDHI VERMA[1] and MANDEEP KAUR[2]

[1]*Research Scholar, University School of Financial Studies, GNDU, Amritsar, Punjab, India*

[2]*Professor, University School of Financial Studies, GNDU, Amritsar, Punjab, India*

ABSTRACT

Since the last few years have led to dramatic changes, today, the customers are in the world where "the use of multiple channels is the rule rather than the exception." This chapter provides a conceptual model to show the relationship between customer experience dimension and loyalty, with multichannel as a moderator, which contributes both to academic research as well as has managerial implications. In this chapter, the author has examined various research papers to examine the basic dimensions of customer experience, and its importance in the present scenario. The chapter consolidates literature to capture the relation between dimensions of experience and loyalty, in the context of multichannel banking specifically. The dimension proposed by Schmitt's (1999a) has been found as the most widely accepted and used by the scholars in their study. The chapter also added to knowledge and understanding, how multichannel banking has the role to influence the relation between experience and loyalty.

7.1 INTRODUCTION

Banking is the sector where technology has a big role to play. The customers have numerous channels to do banking; it can be branch-banking, website, mobile applications, POS, or ATM's, etc. Multichannel banking is the banking of now, and these are the banks that along with the traditional distribution channels, also uses the internet and mobile channels (Hernando, 2007). The experience of the customer thus, does not depend upon anyone channel, but on the overall experience received from all the channels together. With the advancement in technology on one side and an increase in competition on another side, experience has gained a lot more attention and also has been identified as the fourth economic offering after goods, services, and commodities (Horizons, 2016). To define, in the words of Pine and Gilmore, Marketing Pioneer of Customer Experience, as a service or product encounter that "a customer finds unique, memorable, and sustainable over time." In simple words, it can be said as the long-lasting impression formed by customers of the events encountered by them at the time of buying any product or services, etc., (Gronholdt, Martensen, Jorgensen, and Jensen, 2015). It's an area which can help companies to develop a sustainable differentiation, as the experience provided by one company can hardly be imitated by another company. Moreover, commodities, products, and services are no more the preference of the customers, but rather it's the unique and special offerings by the organizations that customers remember for a longer time, and these are derived by the customer experience (Ali and Omar, 2014). Customer Experience Dimensions has been proposed by various authors, but the dimensions provided by Schmitt (cognitive, affective, behavioral, sensory, and social), has been mostly used by the researchers. Scholars have started considering, that providing experiences which are engaging and creates a long-lasting impression in customer's mind are intangible assets, thus is highly valuable in creating loyal customers and adds to the organization's value (Mascarenhas et al., 2006). Thus, the businesses need to focus on creating a positive customer experience to differentiate from the competitors and create superior customer loyalty (Chahal and Dutta, 2015). Jacoby and Kyner (1973) have defined loyalty as "the biased (i.e., non-random) behavioral response (i.e., purchase) expressed over time by some decision-making unit with respect to one or more alternative brands out of a set of such brands is a function of psychological (decision making, evaluative) processes (Brun, Rajaobelina, Ricard, and Berthiaume, 2017)." In this paper, we will discuss about what the literature considers about the various dimensions of customer experience, and regarding its impact on loyalty, while taking the multichannel banking as the central focus.

7.2 OBJECTIVE

The objective of this paper is twofold. The first is to conduct an extensive literature study on the relationship between customer experience and loyalty, while also considering the role of multichannel banking in the Indian Context. The second is to develop a conceptual model on the basis of existing research, in order to identify the relation between various dimensions of customer experience and loyalty, and to check the moderating impact of the channel.

7.3 RESEARCH METHODOLOGY

The conceptual paper provides a comprehensive review of the previous studies that examined customer experience and customer loyalty in the banking industry and other sectors as well.

7.3.1 CONCEPTUAL MODEL (Figure 7.1)

On the basis of existing literature and also supported by Brun, Rajaobelina, Ricard, Berthiaume, & Ricard (2017), the following conceptual model has been developed. The model basically considers the relation between experience dimensions (identified by Schmitt, 1999) and Customer Loyalty. Moreover, it also contemplates the role of multichannel banking as a moderator between this relation.

7.3.2 CUSTOMER EXPERIENCE

Carbone and Haeckel, the pioneer of the concept of Customer Experience, defined it as a lifelong impression of the events that people encounter at the time of buying products, services, and business-a perception produced when humans consolidate sensory information (Hu, Huang, Gilmore, and Tynan, 2017). On the other hand, Schmitt emphasized that CE "provides sensory, emotional, cognitive, behavioral, and relational values that replace functional values" (Gronholdt et al., 2015). From the above definition, we can say that the experience is created not only by the elements which are in the control of the retailer but also by the elements which are beyond the retailer's purview (Brun et al., 2017).

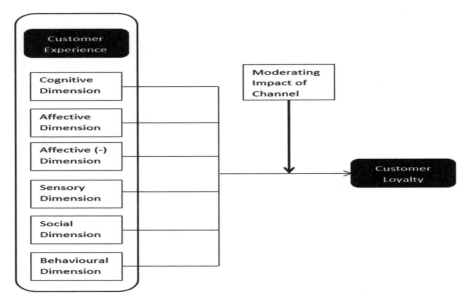

FIGURE 7.1 Conceptual model.
Source: Based on a review of the literature and supported by Brun et al. (2017).

In simple words, when a company intentionally uses services as the stage and the goods as props, to engage individual senses in such a manner that it creates long lasting memory, it is known as Customer Experience (Horizons, 2016). In the literature, there is a varied number of conceptual as well as empirical attempts made to define and identify the main dimensions of customer experience. But, as per an in-depth review of literature, the dimensions identified by Schmitt (Cognitive, Affective, Behavioral, Social, and Sensory) offers a holistic and broad vision of the concept (Brakus et al., 2009; Iglesias, Singh, and Batista-Forguet, 2011; Sahin, Zehir, and Kitapci, 2011; Tsuar et al., 2006; Gentile, Milano, Noci, and Milano, 2007) and have been accepted widely (Lemon and Verhoef, 2016; Brun et al., 2017). The dimensions will be discussed in detail in the following subsections.

7.3.2.1 COGNITIVE DIMENSION

The intellectual experience helps customers shake off their boredom and concerns about the mental process-taking place in the consumer's mind

(Horizons, 2016; Rajaobelina, Brun, Prom Tep, and Arcand, 2018). The dimension considers how customers get immerse in the activity, as it provides the customers with enhanced knowledge as well as enriched relationship quality. The knowledge or abundance of useful information available with customers helps them make well-planned purchase decisions, which leads to an increase in the level of trust. Results of various studies have shown that the cognitive dimension is the most important factor affecting loyalty, be it on any channel (Brun et al., 2017).

The studies found that there is a direct impact of cognitive dimension on various behavioral outcomes, basically brand equity, satisfaction, word of mouth, and trust, and proved to be the most positively evaluated experiential dimension (Rajaobelina et al., 2018).

7.3.2.2 AFFECTIVE DIMENSION

The dimension caters to the feelings and emotions of customers while they encounter the service. The feelings like joy, happiness, surprise, serenity, sadness, anger, fear, disgust, and so on, are all part of the affective dimension; though it constitutes both negative as well as a positive feeling, many of the researchers have only concentrated on the positive affective dimension (Chahal and Dutta, 2015; Rajaobelina et al., 2018). Organizations can enhance their total customer experience by managing these emotions and feelings, along with the quality of goods and services. For instance, empathy by personnel, listening to the queries or solving the problems on time; providing sensory mobile applications that are appealing to eyes and appealing to touch; providing the customer with well-established sequence and clear rules and regulations by the banks or financial institutions, can help enhance the affective experience of the customers and they can take away good impression along with them (Chahal and Dutta, 2015).

Leelakulthanit and Hongcharu (2011) demonstrated that positive feelings have a greater influence in the online services, which further implies that negative feelings have its influence in the offline sector, given the ready access to the personnel available onsite (Brun et al., 2017). In another study, it has been concluded that it is the most important factor to be considered by the banking sector to retain customers for the long term and has the highest influence on loyalty as compared to the other dimensions (Brun et al., 2017; Chahal and Dutta, 2015).

7.3.2.3 SENSORY DIMENSION

Experience is the consolidation of the sensory information received by the customer, in the course of encountering the product or service. The impression formed by the customers from all the five senses (i.e., hearing, touch, sight, taste, and smell) is basically what constitutes the sensory dimension. The relevance of this dimension depends upon the kind of industry; for instance, when a customer is visiting a doctor for some medical problem, the sensory dimension is not that relevant as much others like cognitive and behavioral will matter; whereas, in the retail and entertainment industry, this dimension is of high relevance (Chahal and Dutta, 2015). By using the perfect blend of all the hedonic elements, for all the five senses, the industries can take the experience to all the new levels. The dimension is more workable in the physical settings than the online, as three (touch, taste, and smell) out of the five senses will not work in a virtual environment (Brun et al., 2017).

From the studies, it has been found that the sensory dimension does not significantly impact loyalty, whereas another paper revealed that the dimension has a significant impact on brand equity, satisfaction, word of mouth, and loyalty. The dimension is found to have a significant impact in the web-based environment in the tourism industry, whereas no such impact is found in the banking sector (Brun et al., 2017; Chahal and Dutta, 2015).

7.3.2.4 SOCIAL DIMENSION

In today's time, the experience has become more social than before, as there are numerous feedback forums, chat supports, media, and other touchpoints in the online setting, in addition to the agency setting. Many authors put forth that social dimension has a significant influence on the consumers' purchase behavior, word of mouth, and switching/ retention intention; thus, the authors have considered this dimension to be significant in creating superior customer experience (Brun et al., 2017). The interaction between 'customer and employees' and 'customers and customers' has a significant role in creating a customer experience. It's the dimension which considers the relation of customers with the bank personnel and the other customers as well. Giving due care to the customer's complaint, proper and courteous interaction with customers, sense of belongingness are all that comprise the social dimension (Chahal and Dutta, 2015).

From the literature, it has been found that the social dimension does not have a significant impact on trust as well as on e-trust; moreover, it possesses the lowest scores as compared to the other dimensions on the scale of customer experience. On the other hand, lack of social presence has its impact, albeit negative, on the consumer e-trust (Rajaobelina et al., 2018). Though in the digital era, customers have no. of new e-touch points, but in the traditional non-electronic mode, the employees and customers play a significant role in creating a positive service experience, as compared to the digital mode, where the customer is sitting alone in front of his computer/mobile/tablet screen, etc. Thus in light of the above, the dimension has its impact on loyalty, in the case of agency settings more than the online ones (Brun et al., 2017).

7.3.2.5 BEHAVIORAL DIMENSION

The caring attitude, prompt service, special attention to the customer's time, speedy service processes, etc., can generate a positive behavioral experience, which influences the customer's choice of bank for availing various services (Chahal and Dutta, 2015). The dimension doesn't have a consensus with regard to its inclusion and definition; different authors define it albeit differently. Where one says that the dimension supposes that customers are affected by the customer experience in the form of modifying their lifestyle, habits, and action taken; the other includes the amount of money and time spent in the store to be the part of the behavioral experience (Brun et al., 2017). Now, a day's organizations, give the customers opportunity to be the active participant in manufacturing or designing the product or service, this co-production can reduce the perceived uncertainty of the customers thus making it the component of the behavioral experience, as such this dimension include the behavior related aspects of consumption (Brun et al., 2017; Rajaobelina et al., 2018).

Various studies have tested the impact of behavioral dimension on the customer outcomes, which lead to the conclusion that it doesn't have a significant impact on either trust or commitment; in yet another research, it has been concluded, that the behavioral dimension has a negative impact on loyalty in online as well as offline services, though it is greater in the offline context than in the online. This said, it's very difficult to find the impact of behavioral dimension in the banking context, as it is not readily observable like in other industries as entertainment, whereby you can see the customers getting engaged in the performances, music, etc. Thus it neither satisfies nor dissatisfies the customers; this is how it can be predicted by the negative impact. In the airline passenger context, this dimension has topped all other dimensions to have the highest impact on the customer experience (Brun et al., 2017; Chahal and Dutta, 2015; Rajaobelina et al., 2018).

7.3.3 CUSTOMER LOYALTY

Oliver has defined it as "A deeply held commitment to re-buy or re-patronize a preferred product/service consistently in the future, thereby causing repetitive same brand purchasing, despite situational influences and marketing efforts having the potential to cause switching behavior," in simple words, loyalty is the repeated purchase from the same company.

Satisfaction has a great influence on loyalty; as loyalty in itself is a result of repeated customer satisfaction. Many authors put forth that in order to enhance satisfaction and loyalty, it is significant to concentrate on the customer experience (Bujisic, 2014). Till date, there are limited no. of studies concentrating on internet banking, and the researchers who have studied satisfaction and loyalty in internet banking have mostly neglected the customer experience; they have either taken service quality or satisfaction as an antecedent of loyalty (Mbama and Ezepue, 2016).

In today's intense competition, satisfied customers can easily defect, thus merely having a satisfied customer is not enough, as it's the loyal customer that help companies to strive the competition, and it is the experience which makes the loyal customers, that's why now the researchers have started recognizing experience as the utmost antecedent of loyalty, instead of the previous most searched 'customer satisfaction' (Fatma, 2017; Gupta and Mumbai, 2017). Thus, there is a need to check the impact of positive experiences on loyalty (Ju, Schaffner, Windler, and Maklan, 2012).

7.3.4 MULTICHANNEL BANKING

The emergence of technologies has provided customers with a number of convenient sources to transact with the bank; it can be the bank's employee, website, mobile application, ATM, POS Machines, etc., to cite a few. In simple words, the availability with the bank of more than two channels for providing its services is known as multichannel banking (Hernando, 2007). With the use of such technologies, the service providers reduce their personnel costs and better connection with the customers. On the other hand, the customers, in this way, reap the benefits of reduced time, cost, and energy. The multiple touchpoints have poised consumers with complex and varied kinds of experiences. When a customer uses more than one channel, s/he doesn't isolate one channel from another; thus, there arises the need to study the impact of the experience of one channel on another (Arasu and Manickavasagam, 2011; Eriksson and Nilsson, 2007). The studies revealed that online banking works as the complementary of traditional banking rather than being supplementary to it, as the customers are not at any of the extremes; they consider having a mix of traditional as well as online banking, which is why multichannel banking is considered as the Banking of new generation (Fernández-Sabiote and Román, 2016; Hernando, 2007).

As noted above, nowadays, the techno-savvy customer is surrounded by varied options to transact with the bank and seeks to have an enriched experience. Thus, it's important for the banks to engage customer emotions in a way to provide them with an enticing and long-lasting memory. Therefore, it's significant to grab the customer experience concept across the Banking touchpoints and check how it can lead to develop loyal customers. Despite the success of multichannel model, which considers the issues like channel selection, and motivations, there is a dearth of literature considering how the channels work together or how does it relate to experience and the behavioral outcomes (Fernández-Sabiote and Román, 2016). In the era of multichannel banking, one cannot isolate the channels from one another, as the experience depends on the total experience of all the channels the customer interacts with. Thus there is a need to check the impact of customers' experience of one channel on another (Arasu and Manickavasagam, 2011; Eriksson and Nilsson, 2007).

7.3.5 FINDINGS

On the basis of the extensive literature review, we can say that the area of customer experience has started getting a lot of attention due to the transition from the selling economy to the experience economy. The paper has made a contribution to the literature. Firstly, it helps to highlight that loyalty is the utmost consequence of customer experience, and the relationship between experience and loyalty is stronger than that of satisfaction and loyalty. Secondly, it also helps conclude that each dimension has its impact on loyalty, though the degree of significance may vary. Thirdly, with the multichannel banking in focus, the paper revealed that the impact of customer experience on loyalty moderates on the basis of the channel used. The impact of dimensions basically, the behavioral, sensory, and social dimension vary on the basis of banking channel, as the behavioral and sensory dimension have an impact in the traditional banking mode on loyalty and not in the online context, on the other hand, sensory dimension has an impact in the online setting rather than the offline one. At the same time, the affective (negative) and cognitive dimension does not differ significantly on the basis of channel usage.

7.3.6 IMPLICATIONS

Our findings have managerial implications as well, which suggests that the bank managers have to carefully stage the experience to customers through every channel. They need to make strategies differently for the dimensions, such as behavioral, sensory, and social; as their impact varies on the basis of the channel used. In order to enhance the customer experience through the behavioral aspect, the managers have to be careful that personnel are trained to be courteous, understandable, and shall also make the customer feel a sense of belongingness, to enhance the experience in the agency settings, and on the other hand in the online context, they need to handle the chat groups, the media management in the way to provide information in a chivalrous way. The careful design of the interface aesthetics like the presentation quality, design, information architecture, soft functionality, and easy understandability of language on all the banking channels, will help them to enhance the usage of the channels and which will then lead to the customer's loyalty. Whereas for the elements of cognitive and affective dimension, similar strategies could be formed irrespective of the channel, as channel usage doesn't moderate the impact on loyalty.

KEYWORDS

- **behavioral dimension**
- **customer experience**
- **loyalty**
- **multichannel banking**
- **Schmitt's dimensions**
- **techno-savvy customer**

REFERENCES

Ali, F., & Omar, R., (2014). Determinants of customer experience and resulting satisfaction and revisit intentions: PLS-SEM approach towards Malaysian resort hotels. *Asia-Pacific Journal of Innovation in Hospitality and Tourism (APJIHT)*, *3*(2), 10. https://doi.org/10.7603/s40930-014-0010-2.

Arasu, B. S., & Manickavasagam, S., (2011). *Role of Existing Channels on Customer Adoption of New Channels: A Case of ATM and Internet Banking*. https://doi.org/10.1002/j.1681-4835.2011.tb00317.x.

Brun, I., Rajaobelina, L., Ricard, L., & Berthiaume, B., (2017). Impact of customer experience on loyalty: A multichannel examination. *Service Industries Journal*, *37*(5/6), 317–340. https://doi.org/10.1080/02642069.2017.1322959.

Bujisic, M., (2014). Antecedents and consequences of customer activities. *Management*, 1–8.

Chahal, H., & Dutta, K., (2015). Measurement and impact of customer experience in banking sector. *Decision*, *42*(1), 57–70. https://doi.org/10.1007/s40622-014-0069-6.

Eriksson, K., & Nilsson, D., (2007). Determinants of the continued use of self-service technology: The case of internet banking. *Technovation*, *27*(4), 159–167. https://doi.org/10.1016/j.technovation.2006.11.001.

Fatma, S., (2017). *Antecedents and Consequences of Customer Experience Management: A Literature Review and Research Agenda*.

Fernández-Sabiote, E., & Román, S., (2016). The multichannel customer's service experience: Building satisfaction and trust. *Service Business*, *10*(2), 423–445. https://doi.org/10.1007/s11628-015-0276-z.

Gentile, C., Milano, P., Noci, G., & Milano, P., (2007). *How to Sustain the Customer Experience: An Overview of Experience Components that Co-Create Value with the Customer*, *25*(5), 395–410. https://doi.org/10.1016/j.emj.2007.08.005.

Grace, D., & Cass, A. O., (2004). *Examining Service Experiences and Evaluations*, *18*(6), 450–461. https://doi.org/10.1108/08876040410557230.

Gronholdt, L., Martensen, A., Jorgensen, S., & Jensen, P., (2015). *Customer Experience Management and Business Performance*.

Gupta, R. K., & Mumbai, N., (2017). A study of customer experience about banking services in select co-operative. *International Journal of Management Studies*, *4*, 90–99.

Hernando, I., (2007). *Is the Internet Delivery Channel Changing Banks Performance? The Case of Spanish Banks*, *31*, 1083–1099. https://doi.org/10.1016/j.jbankfin.2006.10.011.

Horizons, S., (2016). In: Joseph, P. II. B., & James H. G., (eds.), *Welcome to the Experience Economy*.

Hu, C., Huang, L., Gilmore, J., & Tynan, C., (2017). *Measuring Gen-Y Customer Experience in the Banking Sector*, 1142.

Ju, U., Schaffner, D., Windler, K., & Maklan, S., (2012). *Customer Service Experiences*. https://doi.org/10.1108/03090561311306769.

Lemon, K. N., & Verhoef, P. C., (2016). Understanding customer experience throughout the customer journey. *Journal of Marketing*, *80*(6), 69–96. https://doi.org/10.1509/jm.15.0420.

Mascarenhas, O. A., Kesavan, R., Bernacchi, M., Mascarenhas, O. A., Kesavan, R., & Bernacchi, M., (2006). *Lasting Customer Loyalty: A Total Customer Experience Approach*. https://doi.org/10.1108/07363760610712939.

Mbama, C. I., & Ezepue, P. O., (2016). Article information: Digital banking, customer experience and bank financial performance: UK customers' perceptions. *International Journal of Bank Marketing*.

Rajaobelina, L., Brun, I., Prom, T. S., & Arcand, M., (2018). Towards a better understanding of mobile banking: The impact of customer experience on trust and commitment. *Journal of Financial Services Marketing, 23*(3/4), 141–152. https://doi.org/10.1057/s41264-018-0051-z.

Schmitt, B. (1999). Experiential Marketing. *Journal of Marketing Management, 15*(1–3), 53–67. https://doi.org/http://dx.doi.org/10.1362/026725799784870496.

CHAPTER 8

Center of Main Interest (COMI): Perspectives and Challenges

ADITYA TOMER,[1] SUMITRA SINGH,[2] and ABHISHEK ROHATGI[3]

[1]Additional Director/Joint Head Amity Law School, Noida, Uttar Pradesh, India

[2]Assistant Professor, Amity Law School, Noida, Uttar Pradesh, India

[3]Scholar, Amity Law School, Noida, Uttar Pradesh, India

ABSTRACT

Insolvency is a situation wherein an entity is unable to pay-off its financial obligations when and as they come due (Definition of Insolvency, Merriam-Webster's Website, 2019). Insolvency is of two types-it may be either "income insolvency" or "accounting report insolvency." Income insolvency occurs when an entity is unable to pay off its dues when they are pending, because of a lack of liquidity (i.e., liquidity crunch). Balance sheet insolvency, on the other hand, occurs when the organization's benefits are not exactly their and liabilities are more (Debt Organization, 2019). Further, to determine whether the insolvency is income insolvency or accounting report insolvency, although out from the ambit of this submission, one may allude to sect 123 of the Insolvency Act (1986) (BNY Corporate Trustees Services Limited and Ors Case Law, 2007).

Before we progress further with this submission, we must first understand the difference between insolvency and bankruptcy. While insolvency refers to a condition of financial misery, bankruptcy refers to the court decision that explains that how the ruined entity facing financial distress is supposed to pay off its debts and comes into the picture when all other alternative methods for the revival of the entity have failed. It is possible that an entity may be insolvent but not bankrupt at any given point in time. However, a constant state of insolvency may lead to bankruptcy. Insolvency can be understood as a step preceding bankruptcy (Financial Ruination Definition, Investopedia, 2019).

8.1 INTRODUCTION TO CROSS-BORDER INSOLVENCY

Cross-border indebtedness incorporates cases where the account holder claims resources in more than one country or where at least one of the loan grantors are not from the country where the bankruptcy procedures are occurring (UNCITRAL Model Law on Cross-Fringe Indebtedness with Manual for Order and Elucidation, 2014). The development of Global exchange and Investment has prompted a flood in the cases of Cross-Border pauper does. In any case, household indebtedness laws have neglected to keep pace with globalization and the International patterns, along these lines, need cooperative energy, and they are frequently sick prepared and need teeth to deal with the possibilities of a cross-fringe nature. This frequently prompts legitimate methodologies which are out of cooperative energy or lacking, which hampers the salvage of undertakings confronting money related issues, and which are not favorable enough to appropriate organization of cross-outskirt bankruptcies, obstructs the protection of the advantages of the wiped-out account holder against squandering, and frustrates streamlining of the estimation of benefits of the indebted person. Further, since there is a shortage of consistency in the dealings with indebtedness instances of Cross-Border nature, there is an atmosphere of distrust that surrounds Cross Border transactions, which in turn impedes capital flow in the International market and disincentivize Cross Border investments and transactions, thus hampering the International flow of Goods and Services (UNCITRAL Model Law on Cross-Fringe Indebtedness with Manual for Order and Elucidation, 2014).

8.2 CURRENT SCENARIO OF INSOLVENCY PROCEEDINGS IN INDIA: A STATISTICAL OVERVIEW

As indicated by the information of The World Bank, the normal time taken to remove pauperdom in India is 4.3 years. Time to resolve insolvency is the duration in terms of years that starts from the filing for insolvency in court (or tribunal) until the resolution of the distressed venture/entity. The time to resolve insolvency is not only an indicator of the fitness of a countries' economic health, but it also a reflection of the fitness of the judicial system and the Insolvency laws that exist in that country. Below mentioned is the normal time taken by the countries in the Indian Sub-Continent to resolve cases of pauperdom:

- Bangladesh: 4 years;
- China: 1.7 years;

- Maldives: 1.5 years;
- Myanmar: 5 years;
- Nepal: 2 years;
- Pakistan: 2.6 years;
- Sri Lanka: 1.7 years.

It is abundantly clear from the abovementioned data that except for Myanmar, all other countries in the Indian sub-continent have a better average time to resolve insolvency as compared to India (Wdi.worldbank.org, 2019). This comparison, although not strictly pertaining to cross outskirt pauperdom, is a testament of the lacunae present in this current regime of the Indian insolvency laws (Data.worldbank.org., 2019).

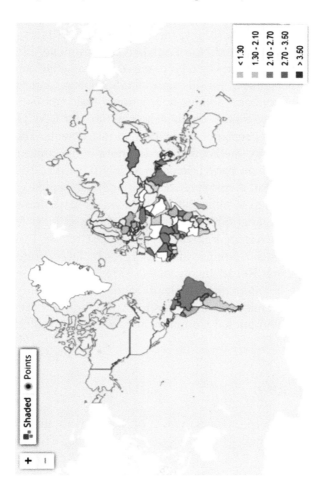

8.3 UNCITRAL

Due to development in technology, the transfer of goods and services is taking place between countries that had no past trade relations. This has led to a situation where all the economies of the world are interdependent and give rise to a global economy. Despite the emergence of this global interdependent economy, the laws that govern the trade in countries around the globe vary from each other according to the jurisdiction. In order to bring uniformity and coordination between various nations that are part of this interdependent global economy, The United Nations Commission on International Trade Law (UNCITRAL) was established by the UN General Assembly on 17 December 1966. This organization of the UN tries to bring Coordination and Synergy between its member states by getting ready and advancing the selection of different authoritative writings, for example, Model laws, Conventions, and Legislative Guides, and Non-Legislative Instruments like Contractual principles that can be incorporated into universal exchange contracts.[1] Apart from the member states, various non-member states, inter-governmental bodies, non-governmental bodies, international and regional organizations are also invited to give their inputs. These efforts by UNCITRAL help in the wide and global acceptance of the drafted instruments. These efforts are further strengthened by the fact that the decisions of the committee are based on consensus and not on votes (Facts About UNICITRAL, 2004).

Originally in 1966, there were 29 member states, which were increased in 1973 to 36 and then to 60 in 2004. The member states are such that they represent different geographic regions, financial, and legal frameworks of the world (UNCITRAL.un.org., 2019).

Some significant writings of UNCITRAL incorporate-UNCITRAL Model Law on Cross Border Insolvency (1977); United Nations Convention on Contracts for the International Sale of Goods (Vienna, 1980); UNCITRAL Model Law on International Commercial Arbitration (1985); UNCITRAL Model Law on International Commercial Conciliation (2002; Facts About UNICITRAL, 2004).

8.4 BACKGROUND

The Ministry of Corp. Affairs, Govt. of India, with its notification no. 35/14/2017, constituted The Insolvency Committee. The purpose of the

---

[1] Basic facts about the international trade law commission by the UN, a guide for UNCITRAL (last visited 4 August 2019).

committee is to review the suggestions and recommendations received from various stakeholders for improvements in the various processes prescribed by The Insolvency and Bankruptcy Code (2016) (Ministry of Corp., 2017). The committee's first report dated March 2018 pointed out the inadequacy in the existing Cross Border Insolvency framework of the I&B Code (2016), referring to it being 'fragmented, complicated, and not at par with global standards' (Ministry of Corp., 2019).

Subsequently, owing to the inadequacy of the existing framework, the Committee suggested the assuming of the UNCITRAL model law on Cross Border pauperdom, which could be added to the existing I&B Code (2016) as a separate chapter altogether. Owing to the technicality of the matter at hand and the need of extensive and comprehensive study to adopt the UNCITRAL model law, the Committee thought it better to submit its recommendation in a separate report (Ministry of Corp., 2019). The Second Report of the Insolvency Law Comm. which deliberated on the reception of the UNCITRAL Model Law on Cross-Border Insolvency, presented its report on 16 October 2018 (Ministry of Corp., 2018).

In view of the point-by-point investigation of the UNCITRAL Model Law, the Committee arranged its Report, which prescribed reception of the Model Law with specific revisions; the proposed draft was attached alongside the Report and named as "draft Part Z" (Ministry of Corp., 2018).

Some reasons to advocate for better cross-boundary insolvency laws are:

1. **Higher Foreign Investment:** The remedy available to foreign creditors under the present code (i.e., I&B Code, 2016) is not elaborate and sophisticated enough to inspire confidence in them that their investments and claims thereunder will be dealt with efficiently and effectively in cases of Insolvency. Selection of the UNCITRAL Model Law on Cross-Border Insolvency will help fill the holes existing in the present enactment and move trust in them. Selection of the Model law will prompt extra recourses for acknowledgment of outside indebtedness procedures, support cooperation, and correspondence among local and remote courts, bankruptcy experts, and the gatherings to the debate. The reception of the Model Law will adjust India to the most generally acknowledged practices in insolvency goals and liquidation. This will prompt more prominent acknowledgment of the Indian Economy by worldwide financial specialists, governments, universal credit associations, and multinational corporations.
2. **Flexibility:** The model law will provide much-needed flexibility, which is needed for the reconciliation of differences with respect

to various domestic insolvency laws. The reception of the Model Law will encourage the cutting out of the important changes in the local indebtedness laws to keep up collaboration with the Global Framework and best practices in the field of universal pauperdom.

3. **Mechanism for Participation:** The presentation of the Model Law will prompt the joining of a powerful component for collaboration and coordination among courts and indebtedness experts, in both outside and domestic locales. This would likewise encourage quicker, amicable, synergetic, and powerful leaders of simultaneous bankruptcy procedures in different locales (Ministry of Corp., 2018).

8.5 MEANING OF COMI-CENTRE OF MAIN INTEREST

Centre of Main Interest (i.e., COMI) can be identified as one of the pillars of Cross-Boundary Insolvency Proceedings in the Model Law on which the "draft Part Z" has been based. The term "Centre of Main Interest" has not been assigned any definition under The UNCITRAL Model Law on Cross-Border Insolvency (Ministry of Corp., 2018). However, in layman terms, Centre of Main Interest can be defined as—"The jurisdiction with which a person or company is most closely associated for the purposes of cross-border insolvency proceedings" (Centre of Main Interests (COMI), 2019). Virgos Schmit Report on the Convention of Insolvency Proceedings provides that the concept of "Centre of main interests" must be construed as the place where the debtor conducts the administration of his interests on a regular basis and is therefore ascertainable by third parties (Miguel VIRGOS and Etienne SCHMIT, 1996).

The genesis of this rule is in the fact that insolvency is a foreseeable risk. Therefore, it becomes important that the jurisdiction for the insolvency proceedings be in a place known to the debtor's likely creditors. This enables the creditors and other interested parties to gauge the legal and financial risks, which would have to be assumed in the case of insolvency to be ascertained and evaluated (Miguel VIRGOS and Etienne SCHMIT, 1996). It also helps in avoiding conflict of laws and jurisdictions and a state of uncertainty.

By using the term "interests" in the phrase "Centre of Main Interest," the objective was to encompass not only commercial, industrial, or professional activities, but also general economic activities, to include the activities of private individuals (e.g., consumers). Further, the expression "main" has been used deliberately to serves as a touchstone for the cases where these interests include activities of varied nature being run from different geographic locations or centers (Miguel VIRGOS and Etienne SCHMIT, 1996).

Since there is no crystallized definition of COMI, the Insolvency Professionals and other interested parties must obtain guidance from the decisions pertaining to insolvency procedures occurring in remote wards and the Insolvency statutes enacted by foreign states having their genesis in the UNCITRAL Model Law as to ascertain what other determinants are to be factored in a while determining COMI.[2]

8.6 PRESUMPTION OF COMI

The model law gives a rebuttable assumption in Workmanship. 16(3) with respect to COMI in the accompanying words:

"Without evidence in actuality, the borrower's enlisted office, or constant living arrangement on account of an individual, is dared to be the focal point of the indebted person's main interests" (UNCITRAL Model Law on Cross-Outskirt Indebtedness with Manual for Sanctioning and Elucidation, 2014).

A similar presumption is found in the proposed Draft Part Z in Clause 14.

This assumption guarantees quick and comfort of evidence in straightforward situations where no contention with respect to COMI is included or expected (11th Multinational Judicial Colloquium UNCITRAL-INSOL-World Bank, 2015). Nonetheless, one drawback to this assumption is that it might prompt maltreatment of the legitimate procedure and gathering Shopping in specific cases to the benefit of the borrower and/or to dupe loan bosses.[3] In any case, Statement 14(2) additionally involves a 'think back period' reaching out to a quarter of a year over from the date of documenting of use for the inception of indebtedness procedures, which is absent in the Model Law. This has been done to evade the unnatural birth cycle of equity and the utilization of the assumption for forum shopping.

Further, Clause 14(3) provides that while trying to determine the corporate borrower's focal point of primary interests, the Arbitrating Authority will direct an evaluation of where the corporate indebted person's focal organization happens, and which is promptly ascertainable by outsiders including banks of the corporate account holder.[4]

---

[2] Mondaq.com. Open at: mondaq's website/ [visited on 23 Jun. 2019].
[3] Report of Insolvency Law Committee on cross border insolvency on Ministry of corporate affairs, Government of India.
[4] Report of Insolvency Law Committee on Cross Border Insolvency on Ministry of corporate affairs,

8.7 RAMIFICATIONS OF COMI

The model Law recognizes remote principle continuing and outside non-fundamental procedures dependent on the area of COMI.

The model law defines them in the following terms:
- "Foreign main proceedings" implies a foreign course of action occurring in the State where lies the main interests of the corporate debtor.
- "Foreign non-main proceeding" implies a foreign course of action, excluding the foreign main proceeding, occurring in the State where lies the establishment of the debtor.

The Indian Draft also has analogous provision based on UNCITRAL in the following terms:
- "Foreign main proceedings" implies a foreign course of action occurring in the State where lies the main interests of the corporate debtor.
- "Foreign non-main proceeding" implies a foreign course of action, excluding the foreign main proceeding, occurring in the State where lies the establishment of the debtor.

The difference between the two will lead to different reliefs being granted by the examining officials.

8.8 DETERMINATION OF COMI

The determination of COMI is to be done in accordance of the UNICITRAL model, which has been incorporated by a number of countries such as the USA, UK, Singapore, etc., within their domestic insolvency laws.[5] Since the Concept of COMI has yet not been incorporated in the Indian Insolvency Laws, i.e., The Insolvency and Bankruptcy Code (2016), the development of this concept will be, to a great extent are influenced by the developments taking place in the foreign jurisdictions.

The following are the leading case laws which expound and throw light on the establishment of COMI:

1. **In Re Sphinx, Ltd. (Bankr, 2006) (United States of America)**
   The court of Bankruptcy of the US, for the Southern District of New York, has invented a widely taken up list of COMI determinants-However,

---

Government of India.
[5] UNCITRAL Model law is based on cross-border insolvency with the guidance of Enactment and Interpretations.

caution must be applied against their mechanical implementation: numerous factors, individually or joined, could be pertinent to such an establishment:

i. Debtor's headquarters location.
ii. Location of debtor's manager (which perhaps could be the holding company's headquarters).
iii. Whereabouts of the chief holdings of the defaulter.
iv. Location of most of the defaulter's lenders or a majority of creditors who could possibly be strained by the case.
v. The authority whose rule would be put in an application in major discourses.

2. **Morning Mist Holdings Ltd. vs. Krys[6] (United States of America)**
In this case, the Court held that any relevant activities, including the liquidation activities and administrative functions, may be considered in the COMI Analysis.

3. **Eurofood IFSC Ltd.[7] (European Union)**
One of the Leading case law for the determination of COMI for a Conglomerate of Companies is that of Eurofood IFSC Limited, the judgment of which was pronounced on 2 May 2006 by the European Court of Justice.

In the EU, the Insolvency proceedings are governed by The European Regulation on Insolvency Proceedings (Council Regulation (EC) 1346/2000). The regulation states the following with respect to the determination of COMI.

"…The center of main interests should be the one where the defaulter can conduct the management for his benefit on a regular basis and which is determinable by the third parties" (The European Regulation on Insolvency Proceedings, 2000).

The principal issue, inter alia, was where the registered office of the parent company was in one state and the registered office of its subsidiary is in a different Member State, and the following conditions are provided:

i. The subordinate routinely addresses its interests in a fashion ascertainable by the third parties in the Member State where its office is registered;

---
[6] Morning Mist Holdings Ltd. v. Krys, No. 11–4376 (2d Cir. 2013).
[7] Eurofood IFSC Ltd., C-341/04 (2006) ECR I-03813.

ii. By virtue of its shareholding and power, the parent company can appoint, command, and control its executives, and it does, in fact, command the policy of the subordinate.

Is the COMI located at (a) or (b)?

It was held by the Hon'ble Court that when a defaulter is a secondary firm whose recorded office and that of its parent firm are located in two separate Member States, the brazenness laid down in the second part of Article 3(1) of the regulation, where it stated that the second center of point interests of the subordinate should be contemplated as situated in the Member State where its recorded office is situated, can be refuted only if factors which are both 'objective and ascertainable' by third parties enable it to be acclaimed that areal circumstance persists leading to the subsidiary having COMI in a place different from that where it has its registered office.

For example, that could be the specimen of a firm not fulfilling any business in the territory of the State in which its registered office is located; this situation is that of a letterbox company).

Adversely, when a firm keeps up with its business in the territory of the State where its registered office is located, the mere fact that its economic selections are or can be directed by a parent company in another State is insufficient to refute the presumption relinquished by the Regulation.

4. **Hertz Corp. vs. Friend8 (United States of America)**

In this case, the court propounded the 'Nerve Center' theory for determining COMI.

It was held that the 'Nerve Center' of a corporation would typically be found at the organization's headquarters.

Further elaborating on the concept of 'Nerve Center,' the court observed that the primary place of business is the nerve center from which it radiates out to its components and from which its officers directs, controls, and coordinates all pursuits without regard to locale, in the furtherance of the corporate purpose.

The court, however, conceded the fact that there may not be any ideal test that would please all administrative and purposive benchmark.

The court took the example of an organization, which may cleave its command and coordinating tasks amid officers who work at various locations, making communications over the Internet.

---

[8] Hertz Corp. v. Friend, 559 U.S. 77 (2010).

However, the Court reiterated the fact that their decision nonetheless points towards a sole direction, towards the center of general direction, control, and coordination.

The Court further if there is no need for the courts to try to weigh corporate functions, assets, or e=revenues distinct in kind, one from another.

This perspective provides a sensible trial, which is comparatively simpler to put into the application; however, it isn't a trial that will generate an output all the time.

5. **In Re Loy[9] (United States of America)**

Nonetheless, in an individual's case, the debtor's constant residence is assumed, in the unavailability of the proof to the contrary, to be the center of the debtor's principal assets.

Other factors that are competent in cases where the debtor is an individual include:

i. The location of debtor's chief assets;
ii. The address of most of the debtor's lenders;
iii. The authority whose law would be put into application in most disputes.

6. **Re: Zetta Jet Pte Ltd and Others (Asia Aviation Holdings Pte Ltd, Intervener) (2019) (Singapore)**

Re: Zetta Jet Pte Ltd and others (Asia Aviation Holdings Pte Ltd, intervener) (2019)[10] is the latest landmark case law in the arena of Cross Border Insolvency. The UNCITRAL Model Law on Cross-Border Insolvency has been adopted by Singapore on 23 May 2017 by way of the Companies (Amendment) Act (2017) (Companies (Amendment) Act, 2017). In this case, The High Court of Singapore deliberated upon the relevant time for the determination of COMI and the factors to be taken into consideration while determining COMI.

Brief Facts of the Case:

Zetta Jet Pte Ltd is a firm embodied in the laws of Singapore. It is also the sole owner of Zetta Jet USA, Inc, which has been incorporated under the laws of California, USA; both of these companies deal with renting and chartering of aircraft.

Both these companies are a part of a group of companies consisting of 16 entities, which have been incorporated into the British Virgin Islands always.

---

[9] In Re Loy, 380 B.R. 154 (Bankr. E. D. Va. 2007).
[10] Re: Zetta Jet Pte Ltd and others (Asia Aviation Holdings Pte Ltd, intervener) [2019] SGHC 53.

Another Company by the name Asia Aviation Holdings Pte Ltd (AAH) owns a 34% stake in Zetta Jet Singapore.

In 2007, Chapter 11 voluntary bankruptcy proceedings were initiated against Zetta Jet Pte Ltd and Zetta Jet USA in Los Angeles.

Subsequently, AAH instituted an application against Zetta Jet Singapore for breach of the Share Holder's Agreement. In pursuance of the initiation of the abovementioned suit by AAH against Zetta Jet Singapore, an Injunction was ordered against Zetta Singapore from taking any action/ step under the Chapter 11 bankruptcy filings in the Los Angeles Court. Inspire of the abovementioned injunction, Chapter 11 filings were converted to Chapter 7, and a Trustee was appointed thereunder. The trustee so appointed was tasked with the recognition of the proceedings in Singapore. The injunction so ordered by the earlier court was set aside by the High Court of Singapore.

Because of this, the following two questions came to the fore:

i. What is the relevant time for the recognition of COMI?
ii. What factors are to be considered while determining COMI?

Observations and decision by the High Court:

i. **What is the Relevant Time (date) for the Recognition of COMI?** The Model law is silent on this aspect of Cross Border Insolvency. However, the 2013 guide provides that COMI should be determined as at the date of the commencement of the foreign insolvency proceedings.

The court, while considering this question, deliberated on the approaches followed by three Countries vis. USA, Australia, and England.

    a. **The English (European) Position:** The approach adopted by the English Court is similar to that of the 2013 guide. According to Cross-Border Insolvency Regulations 2006 (SI 2006 No 1030) (UK) Sch. 1 and the Recast European Insolvency Regulations (EIR), the debtor's COMI is to be determined on 'the date of the application for the initiation of Insolvency proceedings abroad.'

It was argued that the concept of COMI is used differently by the English Courts. The English Courts use the concept of COMI to determine:
- If the Recast EIR is applicable to the case present before them.
- In which member state is the proceeding being to be initiated.[11]

---

[11] Art 3(1) of the Recast EIR states.

On the other hand, the model law uses the concept of COMI to determine the relief, which is to be granted under the model law in cases of foreign main and non-main proceedings.[12] Owing to the abovementioned reasons, the approach adopted by the English courts was rejected.

b. **The Australian Position:** According to the Australian laws, the debtor's COMI is to be determined at the 'time of the hearing of the recognition application,' but regard may be had to historical facts which led to the position at the time.

The Singapore High Court failed to find any material difference between 'the dates of the application' from 'the date of the hearing.' Hence, the court remarked that the Australian Position is substantially like that of the US.

c. **The US Position:** According to Chapter 15 of the US Bankruptcy Code, the relevant time for determining the COMI is at 'the filing of the application for recognition.' The court held that this approach rule helps in the harmonization of efforts and synergy in cross-border proceedings.

The court finally accepted the US position on this aspect and gave the following reasons for its decision:

- That articles 2(f) and 2(g) of the Singapore Law, which deals with the determination of the foreign main and non-main proceeding, refer to the proceedings that are "taking place." The use of the present tense implies that the proceedings are already underway at the time of ascertaining the debtor's COMI.[13]
- This position would allow the court to account for shifts in the debtor's COMI in the period between the commencement of the foreign insolvency proceeding and the date the recognition application is filed.[14] However, the latter part of the judgment puts restrictions on the shifting of COMI to avoid any kind of misuse.

ii. **What Factors are to be Considered While Determining COMI?**
The High Court held that while considering factors which help determine COMI, that it would take into consideration the weight that a

---
[12] See Art 20(1) of the specimen law.
[13] Similar terminology is used in the "Draft Part Z."
[14] However, a note must be taken that that Indian amendment has introduced a look back period of 3 months, not present under the UNCITRAL specimen law.

creditor would attach to a particular factor, and take into account the factors on which a creditor might hedge his judgment as to whether he would afford credit to the company or not.

The High Court opined that efforts should be made to induce permanence in the principles for determining COMI.

The High Court made a departure from the nerve center theory that has been adopted by the US courts for determining COMI. According to the court, efforts should be made to ascertain the gravity of the importance of the objectively ascertainable factors located in different states. The test should be to determine where the mass rests.

Furthermore, the court held that COMI analysis should not be restricted to the Debtor Company only. The Supreme Court opined that while determining COMI, the factors relating to the whole group of companies, of which the debtor company is a part, should be considered.

The High Court of Singapore considered the following heads important for the determination of COMI:

1. The location from which control and direction were administered. The court considered the location from which control and direction were administered as material for determining the COMI. Further, the court held that while determining COMI, the inquiry by the court should be limited to who is actually administering the operations. Any question as to the legal capacity to administer the operations are to be left open for the appropriate court and tribunal to determine.
2. The location of clients;
3. The location of creditors;
4. The location of employees;
5. The location of operations: The court held that this factor was of much importance while determining COMI. However, in this particular case, the importance of this factor was diminished owing to the below-mentioned reasons. The court took notice of the fact that since Zetta was involved in renting and chartering airplanes, the location of assets and operations would not be of much importance. This reasoning was also hedged on the fact that the operations of the Zetta also included a fair share of international flights. However, the court remarked that they would have held conversely if the transport business would have been primarily domestic inland transport.

6. Dealings with third parties; and communication with key customers, vendors, and creditors have been considered material while determining the COMI. Any communication which is easily perceivable as and ascertainable by third parties goes a long way while determining COMI.
7. The governing law. It was held that this factor was of less importance in most situations now because of the demise of the rule in Gibbs outside England.

    The court also considered whether the location that the foreign representative was operating from could be used to determine COMI?

    The Singapore High Court, in this aspect, deviated from the US approach and held that this factor should not be considered as relevant while determining COMI.

8.9 CONCLUSION

The proposed amendment (Draft Part Z) formulated by the Insolvency Law committee should be adopted at the earliest to bring the domestic insolvency laws of India at par with the global standards. Adoption of these amendments will provide a sense of safety to the foreign investors with respect to their investments, providing them with confidence and prompt them to make more investments in Indian markets and further help in the integration of the domestic Indian economy with the Global economy. However, caution should be adopted while adopting the standards of the UNCITRAL Model Law, as discussed above, since most of the rules of the Model Insolvency Law have been customized according to the needs of the adopting countries. The interpretation of Model Insolvency Law should consider both-The domestic interests and international standards. Synergetic international efforts will also be required for the complete and successful implementation of the rules enshrined under the model law. Ambiguous terms like COMI should be better defined and explained. Efforts should be made to introduce as much clarity as possible, especially in the meaning of COMI, since it is pivotal for all the cross-boundary proceedings that are supposed to take place under the Model Insolvency Law. There should be joint efforts by all the adopting states of the UNICITRAL Model Law to bring about permanence in the determination of key concepts, especially to concepts as fundamental as COMI.

KEYWORDS

- **cross-border insolvency regulations**
- **European Insolvency Regulations**
- **model insolvency law**
- **ramifications**
- **United Nations Commission on International Trade Law**
- **Zetta**

REFERENCES

Virgos, M., Schmitt, E. (1996). *Miguel VIRGOS and Etienne SCHMIT. Report on the Show on Indebtedness Procedures.*

EU Council (2000). *The European Regulation on Insolvency Proceedings, 3*(1).

United Nations Commission on International Trade Law (2004). *Facts About UNICITRAL* (4^h edn.), 1–2. UNCITRAL's https://uncitral.un.org/sites/uncitral.un.org/files/media-documents/uncitral/en/12-57491-guide-to-uncitral-e.pdf.

United Nations Commission on International Trade Law (2014). *UNCITRAL Model Law on Cross-Fringe Indebtedness with Manual for Order and Elucidation*, 8, 19–21.

United Nations Commission on International Trade Law (2015). *11th Multinational Judicial Colloquium UNCITRAL-INSOL-World Bank*, 5.

UNCITRAL-INSOL-World Bank (2017). *Companies (Amendment) Act 2017*, 41.

United Nations Commission on International Trade Law (2019). *Often Posed Inquiries - Command and History | Joined Countries Commission on Universal Exchange Law*. UNCITRAL.un.org, https://uncitral.un.org/en/about/faq/mandate_composition/history (accessed on 21 October 2020).

UNCITRAL-INSOL-World Bank (2019). *World Progression Pointers | The World Bank.* Wdi.worldbank.org, World Bank's website.

Bankr, S. D. N. Y., (2006). In Re SPhinX, Ltd., 351 B.R. 103.

BNY Corporate Trustees Services Limited and Ors Case Law, (2007). United Kingdom.

Centre of Main Interests (COMI), (2019). *UK Practical Law Website*. UK Practical Law. https://uk.practicallaw.thomsonreuters.com/6-503-3605?transitionType=Default&contextData=(sc.Default)&firstPage=true.

Data.worldbank.org. (2019). *Time to Resolve Insolvency (Years) | Data*. [online]

Debt Organization, (2019). *Debt's Organization/Questions/Ruined.*

Definition of Insolvency, Merriam-Webster's Website, (2019). Merriam Webster's Website/Dictionary/Insolvency, https://www.dictionary.com/browse/insolvency.

Ministry of Corp., (2017). *Affairs, Government of India, Committee of Insolvency.*

Ministry of Corp., (2018). *Affairs Government of India, Report of Indebtedness Law* (pp. 5, 6). Board of trustees ON CROSS Outskirt Bankruptcy.

Ministry of Corp., (2018). *Affairs, Government of India, Report of Indebtedness Law* (p. 12). Advisory group on cross Outskirt Indebtedness.

Ministry of Corp., (2018). *Undertakings Legislature of India* (p. 31). Report of indebtedness law board of trustees on cross Fringe indebtedness.

Ministry of Corp., (2019). *Affairs, Government of India, Insolvency Law Report Committee* (p. 5).

Sarkar, S. (2019, October 28). *Digital Payments system in India: Facts, Features, Methods & Best Apps*. Retrieved December 10, 2019, from Sole Blogger. https://www.soleblogger.com/digital-payments-system-in-india/

CHAPTER 9

Digitalizing Business Innovation

RAMAMURTHY VENKATESH

Research Scholar, Faculty of Management, Symbiosis International (Deemed University), Maharashtra, India

ABSTRACT

Why is so much talked about digital and business transformation in the past few years? Almost every business in every industry is disrupted, as evident from what they say in all forms of their business communications. Contemporary studies on business innovation and business transformation studies exhibit diverse approaches and recommendations. There is no doubt that business innovation has become a key factor for strategic and competitive advantage for business growth as agreed by both academic and business worlds. Still, there are gaps in contextualizing digital possibilities and business innovation or, for that matter, business transformation in practice. First of all, is business innovation directly connected to business transformation? Secondly, is digitalization is the only means for business transformation? Such mixed connotations are driving business leaders and researchers to probe beyond the traditional boundaries of strategic management, entrepreneurship, and technology management. In other words, there are emerging domains of business innovation with conflated approaches to linking digital possibilities and business transformation. This book chapter is to address some of those complementary constructs in forming a new perspective of four dimensions for understanding digitalization and business innovation.

Understanding and articulating business models (BMs) as a blueprint of business strategy (BS) is the best way to start with. BM elements then address the primary concern of demystifying the relationship between business innovation and business transformation. Business transformation and digitalization are posing a lot of challenges to these business managers and forcing them to alter the BM components frequently. So, as a second step, certain propositions can be made around digitalization and business innovation as an overlay of BM elements. These propositions shall also distinguish the differences between digitization and digitalization. For better practical application and understanding, inferences of these propositions shall bring out working definitions for business innovation, business transformation, and digitalization. As a third and subsequent step, service BMs shall be contextualized in terms of digitalization and business innovation. Connotations such as connected business networks, the experience economy, and as-a-service BMs are then illustrated in terms of digitalization and business innovation. While talking about business networks, of course, there are new paradigms of platform BMs and ecosystems. In other words, business ecosystems powered by digital platform models have expanded the reach of cross-border investments and eased market expansions, which were major barriers for globalization, a few decades back. So, understanding platform models and ecosystems as a complimentary adaptation of economics and scale and economies of scope are necessitated. In essence, it is logical to extend that digitalization is, in effect bringing about economies of speed to platform models and ecosystems.

Finally, all the connotations around digitalization, digital BMs, platform models, and ecosystems are extended as different paths to business innovation. Thus this chapter will attempt to traverse different paths of digitalization for business innovation to form a new grid model. This new grid model-543 Grid-with three hierarchical dimensions present an integrated view that can be equally applied to different contexts-Large multinational companies or SMEs, product companies or service providers, traditional organizations, or digitally advanced NextGen businesses. This chapter attempts to triangulate the strategic, tactical, and operational imperatives for a hierarchical conceptualization of business innovation and digitalization as a 543 Grid model. This new way of approaching to business innovation shall contribute to further improvisation by business practitioners and academicians.

9.1 BUSINESS INNOVATION CONTEXT

Modern-day businesses are essentially driven by digital capabilities and innovative approach towards value creation and competitive advantage. Business leaders globally are challenged with myriad possibilities of establishing and sustaining business capabilities, irrespective of the size and industry. Innovation is seen as topmost weapons to create, deliver, and capture business value and business growth. However, there are gaps in contextualizing digital possibilities and business innovation or, for that matter, business transformation in practice. First of all, business innovation directly connected to business transformation or not is a question that needs to be addressed. Especially service business models (BMs) are facing challenges due to a lack of clear framework and metrics for practical application and quantifiable measurement of innovation capabilities and digital capabilities. Though there is no dearth of knowledge around innovation and entrepreneurship, a concentrated approach to combing different perspectives of digitalization and business innovation is not yet fully mature. In addition, it is apparent there are no clear innovation metrics or innovation management standards, yet that address how innovation-driven business performance shall be measured and controlled. Thus, this study focuses on arriving at a layered approach to understanding the linkages between digitalization and business innovation. Various contemporary constructs and dimensions of business innovation are reviewed to develop a practical model or framework in terms of innovation sub-construct, digital capabilities, and with practical applicability. Accordingly, we propose a conflated and integrated view of business innovation built over three hierarchical dimensions. Then we proceed to propose a grid model in hierarchical dimensions integrating possible innovation and digitalization constructs in terms of leadership, enablement, and orchestration as they lead to business innovation.

In the current world of dynamic business challenges, it is more often how the business leadership manages and surpasses the expectations of all its stakeholders, who always demand more. The modern business environment is highly driven by new technologies, agile competition, and well-informed customers, and various other factors. This is has created a greater need for all business enterprises to better understand and adapt to their BMs. BM, in its simple terms, describes the business logic or design of how the business enterprise employs value creation, value delivery, and value capture mechanisms to make an economic profit. In more broader terms, the BM describes how well the business enterprise understand customer needs, their willingness and ability to pay for the value in exchange, how it responds to deliver that value to customers, and how it plans to convert those payments to economic profits through design and operation of the various business components in the value chain.

Business innovation is defined and understood in diverse ways, as noted from various business literatures. There are many ontological representations for depicting BM elements. To understand and apply innovation to BMs, it is first essential to understand and express BM as a set of clear working definition. What do business models represent?

- Business models are essentially a blueprint of business strategies to drive business goals.
- Business models initiate the process of viable business plans and ways of financial resources.
- Business models portray the key value propositions and act as a framework for operational planning.
- Business models serve as a tool to measure business outcome and level of competitive advantage.

It is also clear that these BMs cannot be the same forever. Constantly reviewing and innovation of BM is more than imperative for sustainable business. However, the process of BM innovation as a whole poses lots of challenges to business leaders. They need to be creative and insightful for reaping the benefits of business innovation. Here again, business insights are a combined and aggregated intelligence about the competition, customer needs and partners who can support business outcome. Creative leadership is all about using the business insights and enhancing customer value creation and the same time ensuring that it is not easily replicated by competitions.

9.2 CONTEMPORARY THEORIES ON BUSINESS INNOVATION, BUSINESS MODELS (BMS), AND DIGITAL TRANSFORMATION

With multi-disciplinary attention to BM innovation studies, academicians, and industry practitioners are keen to develop measurement models, greater generalizability of results and a greater methodological sophistication (Zott, Amit, and Massa, 2011; Spieth, Schneckenberg, and Matzler, 2016; Clauss, 2017). There are recent evidences of gathering momentum addressing the business innovation metrics, yet only few quantitative empirical studies are available on measurement instruments and innovation frameworks. Thus, a multidimensional approach is necessary to cover the generalizability and development metrics-driven innovation frameworks. Clauss (2017) has attempted to address this research gap and developed a set of new scales to measure BM innovation from value perspective. Till early 2000s, studies on BS, innovation, and entrepreneurship, technology management and business process management (BPM) were more dominant but in the last decade, interest of business researchers have turned to dynamics of business innovation and business transformation (Venkatesh, Mathew, and Singhal, 2018).

Comprehending and conflating the extant literature on BMs and business innovation, we see there is a need to follow a hierarchical approach to the

BM dimensions. We conflated a three-level framework with the first order, second order, and third order dimensions of business innovation. Firstly, we recognize that BM itself to be considered as central to first order business dimensions. Every business starts with a strategy, modeled into a viable business plan; business processes are composed, delegated, and executed by functional teams and finally evaluated and measured for performance goals and improvements. This primary first order dimensions can be described as: Analyze, Model, Process, Execute, Evaluate (AMPEE), and with BM in the centre. These first order dimensions are drawn from the existing widely accepted constructs of BS, business model (BM), BPM, business value chain (BVC), and business performance evaluation (BPE) respectively. It is logical then to move to the second order business innovation dimensions with the sub-constructs as extended from the first order AMPEE dimensions. These sub-constructs draw references from literatures reviewed and many previous studies (Baden-Fuller and Morgan, 2010; Demil and Lecocq, 2010; Zott, Amit, and Massa, 2011; Casadesus-Masanell and Zhu, 2013; Wirtz and Daiser, 2016; Venkatesh and Singhal, 2018). It is evident that current studies on BMs and business innovation are greatly influenced by proven analytical approaches such as Porter's Value Chain, BM Canvas, telemanagement (TM) Forum's Business Process Framework, and Balanced Score Card in an integrated manner. This integrated approach to business innovation best practices will be of significant practical value.

We further extend and conceptualize the second order business dimensions as a combination of four constructs such as value, innovation, digital, and enterprise (VIDE) framework. Obviously, value dimension has found lot of mention in the current academic literature and most of them are very closely related to popular triage of value creation, value delivery, and value capture elements of BM Canvas as conceptualized by Osterwalder, Pigneur, and Tucci (2005). Many recent BM studies also reflect importance of value dimension as principal to measuring innovation (Wirtz and Daiser, 2016; Clauss, 2017; Heikkilä, Bouwman, and Heikkilä, 2018). Relative to value dimension, recent surge in digital perspectives of BM are very interesting. Digital dimensions are mostly seen as trends in digital transformation. We note that business transformation and digital are very closely discussed due to dominance of digital technologies in business world. Particularly strategy management studies talk a lot around digital BS. Exponential increase in digital possibilities and technology advancements may be the reason for such growing interest on digital BMs. This particular inference is supported by the findings of many recent articles. Very notably, Bhardwaj et al. (2013) speaks about of the power of digital capabilities as an enabler of economies of scale, scope, and speed. This notion of digital proliferation as economies of speed is further extended as Digital Matrix (Venkatraman, 2017). In this book, author nicely explains digital matrix model that further lead to emerging concepts of business ecosystems and digital platforms. Therefore, it is apparent that business innovation constructs should also include digital capabilities of the business, as success of today's business largely rely on connected business environment and internet possibilities. Apart from value and digital dimensions, the third view of enterprise dimension is very important from enterprise and organizational perspectives. Studies on enterprise perspectives covering people, process, and systems or technology have long legacy even before BM studies (Euchner and Ganguly, 2014; Frankenberger, Weiblen, Csik, and Gassmann, 2013; Hansen, 2007). With the growing confluence of multidimensional approach to business management, we find contemporary studies on enterprises and organizational studies are more closely related to business innovation (Trimi and Berbegal-Mirabent, 2012; Spithoven et al., 2013; Müller, Buliga, and Voigt, 2018).

From the practitioners' perspective, we see there are lot of overlapping discussions around digital capabilities and business innovation. Terms such as 'business transformation,' 'digital enablement,' 'digital transformation' and digital BMs have taken center stage in every business conference. Most of leading technology solutions and services companies such as Cisco, Accenture, IBM, CapGemini, and others often promote 'Digital business transformation' as their core offering and put lot of importance around innovative business outcome. Particularly in today's service centric business environment, service BMs highly dependent on digital resources (Dobni, 2008; Cavalcante et al., 2011; Heikkilä, Bouwman, and Heikkilä, 2018). Recent article by MIT regarding the trends of digital economy, articulates the challenges faced by new business leadership due to digital disruptions and the obvious need for rapid business transformation. BM is often considered as a blueprint of BS and it needs be innovated for a sustainable business transformation. Figure 9.1 provides a graphical view of how types of innovation and transformation initiatives can be linked to BS and BM as adapted from relevant literature.

The terms innovation and business transformation are often used interchangeably blurring the understanding further. Further conversation around innovation and transformation need a simple and concise understanding of the relations between these terms. Innovation in business terms simply means a set of processes for making something new or doing something in a novel way to add value. Innovation is not just confined to technology or invention. It covers all the stages of ideation, development, implementation, and commercialization. This working definition can be extended to the understanding and relations between BS and BMs. As defined by Gartner, innovation management initiatives focus on disruptive or step changes that transform the business in some significant way. Innovation management concepts very closely relates to strategic management and technology management.

Transformation, on the other hand, describes those ongoing processes that usually happen when there are change in long-term strategic objectives and BS and it takes longer time. Innovation, on the other hand, refers turning ideas into reality. It usually refers to a relatively quicker and creative activities leading to better business benefits. In the current business world, major multinational and global companies are adapting to use innovation and subsequent transformation to benefit both customers and their own corporate strategies.

Innovation matrix is provides a simple yet clear distinction between types of innovation in terms of the problem definition and the efforts to find innovative solutions. Traditionally, it starts with basic research and development and then moves on to breakthrough innovation or disruptive innovations that may bring radical changes to the way industries and market operates. However, most innovation efforts lead to sustainable innovation as business leadership aim to keep doing things better and profitable way most of the time. This in turn leads to improvement in existing capabilities in existing markets needed for business transformation.

Business Transformation is essentially a change management strategy with key focus to aligning people, processes, and technology initiatives of a business enterprise. It is more centered around in the changes in the BS for long term benefits. It may be noted that incremental changes and transitions in business leadership, simple expansion of existing business to other markets or industries are not to be confused with business transformation. In addition, business transformation is not necessarily technology centric or IT focused as it may appeal to some. Reason being, concepts of business transformation gained traction in recent times as promoted by some major IT companies who have successfully re-branded in selling their integrated information systems and consultancy services as business transformation.

Business Transformation matrix is yet another way to understand the types of transformation initiatives. Internal and external focus of the business and their level of intensity can be considered to classify the transformation types as in Figure 9.1. Technology upgradation or service modifications can be considered as internally focused initiatives which may or may not be driven by innovation. Functional or IT transformation mainly focuses on changes in business processes, integration of IT for business processes improvement or automation to put IT resources for better and effective use.

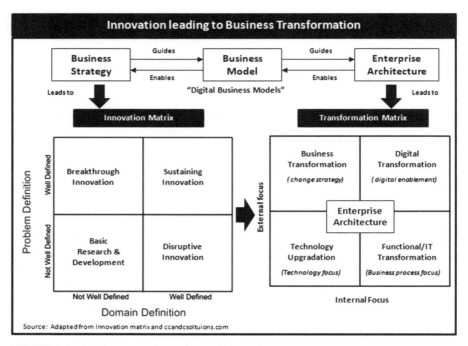

FIGURE 9.1 Business transformation and innovation landscape.

Business transformation or Enterprise transformation is more outward-looking and external driven aligned to long-term strategic objectives (Knight et al., 2005; Wirtz, Schilke, and Ullrich, 2010). As per Harvard Business Review, business transformation can be broken down to three categories such as operational (using newer technologies as long-term solutions), operational model (changes to core activities and business processes) and strategic (radically changing the entire BS) of the firm. However, there are varying degrees of consensus about what is really meant by business transformation. It is vital to understand that efforts such as outsourcing, downsizing, seeing consultancy services, or some limited IT initiatives are not business transformation. Business leaders should understand that business transformation is a must and most important journey that current business enterprises across every industry should undertake for a sustainable future.

9.3 DIGITALIZATION AND DIGITAL TRANSFORMATION

Digital transformation describes business practices and processes used by business enterprises to improve profitability and long term economic benefits by adapting to digital enablement and changes to their core technology. While there is considerable overlap with business transformation, the approach of digital transformation is a clear distinction between innovation and transformation. Digital transformation accelerates the business transformation by fully leveraging the potential opportunities and capabilities of digital technologies. Here, digital enablement is the key and it greatly influences the business transformation journey through digitized business processes for enhanced customer experiences and innovative BMs.

VUCA world-Volatile, Uncertainly, Complex, and Ambiguous is a recent buzzword to describe current business environment. New digital imperatives are providing many new opportunities and challenges in this VUCA world. Customers are more than connected ever, digitally informed in real time, and have adapted to digital culture. Every consumer or user is seeking superior value and customer experience in all their touch points. To meet their demands and expectations, BMs need to embrace digitalization of products and services as it is the only sure way to remain successful and sustainable.

Are these BMs stand fixed once made? Not entirely. Firms may need a portfolio of BMs addressing different markets and products. That is, modern BMs are essentially digitally BMs made of combination of combination of several BMs individually powered by digital capabilities. These blended BMs powered by internet possibilities make up today digital economy built as a network of digital platforms in the value chain. In other words, business transformation driven by digital economy can be accomplished by embracing to a digitization of products and service portfolio. So, what is digitization? Here we need to understand the distinctions among three key concepts-digitization, digitalization, and digital transformation.

Digitization is the simply the way all information is converted to digital format. Once digitized, it is easy and flexible to store, access, process, and share the information. It may also be noted that digitization of information is not restricted to only select business data or select industries. It runs all over the business environment and applies to every industry and all business processes that produce data which can be shared and used by cross-functional teams. Digitization is mainly transferring all the information to digital formats.

Some examples are such as embracing paperless office or digitized workflow management systems.

Digitalization or Process Digitalization means some form of digital technologies are being used to manage the business processes and information digitally. This is to ensure business processes are more effective and efficient with better productivity and profitable business outcome. Main benefits of digitalization are to provide economies of speed to business, with cost optimization, better customer experience, and new revenue streams. Digitized products and services will also offer better ways to co-create value and better monetization of products and services. Digitalization of processes focuses on product and services as well as value creation perspective. It is important to recognize that mere digitization of information is not digitizing products and services. Unfortunately, this is not understood unequivocally and hence when we refer digitization or digitizing products or services, it is actually referred to digitalization. Figure 9.2 provides a consolidated view of digitalization of productions and services.

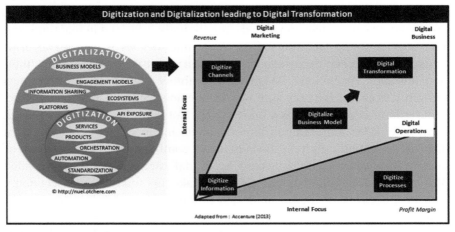

FIGURE 9.2 Digitization of products and services portfolio.
Source: Adapted from Amamoo-Otchere, 2016.

Finally, digital business transformation or simply digital transformation means adapting to digital concepts for novelty in value creation and value delivery. It is more than simple process automation or digitalizing business process. It involves modifying the entire BM and re-designing the value chain for long-term business growth and customer value. This digital business transformation approach should be aligned with BS and ways of creating innovative value propositions. More importantly, digital business transformation efforts should be measured and controlled for value capture and improved stakeholder value (Bouwman, de Reuver, and Nikou, 2017; Bouwman et al., 2018; Venkatesh, Singhal, and Mathew, 2019). Connotations such as connected business networks, experience economy and as-a-service BMs can also be illustrated in terms of digitalization and business innovation. BM canvas elements can be extended to explain the imperatives of digitizing products and services and in turn digitalization of processes which can mature into digital transformation and digital BMs.

Whitelock (2017) provides a detailed commentary about the trends monetizing digital BMs and the conflicting notions of business managers about what they really mean by digitalization of business process or digital transformation. In addition, due to lack of clear distinctions, there are reasons to believe why business transformation and digital transformation is considered one and the same by many (Pandya and Anand, 2008; Ashurst, 2012; Nadeem, Abedin, Cerpa, and Chew, 2018). How do the concepts of digitization, digitalization, and digital transformation relate to each other and how business transformation is can be illustrated as an array of digital transformation efforts? A typical example of how a newspaper company with print only edition can evolve into a digital media company can be useful to understand this context. Inspired by the works of Andersson and Lyckvik (2017) and reports from World Economic Forum (2017) Table 9.1 summarizes the transformation journey of a typical newspaper company to digital BM as components of well-known BM canvas.

9.4 EMERGENCE OF DIGITAL PLATFORMS AND DIGITAL ECOSYSTEMS

Digitization of products and service needs better understanding of the value chain activities and in turn the supply chain efficiencies. Now the question is what is the real difference between value chain and supply chain? Concept of supply chain originates from the product and manufacturing based business thinking while the value chain is promoted in 1980s by Michael Porter as a source of competitive advantage.

Traditionally supply chain essentially refers to all collection of parties such as suppliers, manufacturers, distributors, technology, information systems, transporters, and other support activities needed to manufacture the product that the customers want at the lowest cost possible. Here the focus is on cost reduction and efficient production. Value chain is defined as the combination

TABLE 9.1 Typical Business Transformation of a Print Newspaper to Digital Media Business

	Business Transformation: From Traditional business to Digital business by Digital Transformation			
Traditional Model	**Transition +/– Outsourcing**	**Digitized Model**	**Digitalized Model**	**Digital Transformation**
Shifting focus and business transformation	Focus on cost reduction and operational efficiency; Productivity	Digitize information flow, Communications, use on digital resources for automating processes and workflow	Digitalize value propositions digital products and services new revenue streams	Digital business models, use of ecosystems, multisided network value chain, platform models
Value Proposition				
"Ethical and qualitative journalism, real time news, entertainment, media for marketing and connecting people, social responsibility"				
Daily Newspaper (News, articles, ads)	Limited locations → More locations	Multi-city Print Editions	Print Editions (less focused)	+Digital Mobile App Edition
Weekly Magazine (Features, special stories)	Multi-city reach	Online Edition (limited access)	Online Portal Edition (Freemium)	+Personalized Content
			Wider & Instant reach	+Real Time stream
CRM	Loyal customers + Subscribers	Email	Portal Feedback	+App feedback
Letter to Editor		Websites/Online feedback	Click Stats	+Portal Feedback
Feedback Forms (print)	Long term Corporate relations	MIS reports	MIS/ERP Reports	+BI Reports
Agents Reports		Call Centre	Digital Market Research	+Social Media Analytics
Personal Visits				
Customer Segments				
News/Magazine Readers (for readership)			+ Freemium Readers	+Location Based Marketing
Advertisers/Corp Firms (for ads)			+Market Research firms (usage data)	+Customer Profile data buyers
			+Advertisers/Corp Firms (digital mktg)	+Platform and Network partners
				+Digital media services

Digitalizing Business Innovation 161

TABLE 9.1 (Continued)

Business Transformation: From Traditional business to Digital business by Digital Transformation

Traditional Model	Transition +/- Outsourcing	Digitized Model	Digitalized Model	Digital Transformation
Channels				
Distributors/agents	Distribution management	Website	Online Edition Access (Freemium)	+Digital Mobile App
Newsstands/Bookstores	Subscriber Management		Social Media Presence	+Personalized Streaming
Direct Salesforce (Ad Sales)				
Key Activities				
Journalism & Editorial	IT for internal communication	ERP based Operations	Outsource Printing	-No more Printing
Production Logistics (Inbound)	Advanced Printing Press	Digital Printing	Outsource Distribution	-No print distribution
Printing Operations (Operation)		IT Platform for Distribution mgt		+Customer Analytics
Distribution Logistics (Outbound)		Internet Ads/SEO		+Social Media Engagement
Marketing (Print, Outdoor, advertising)				
Customer Service				
Support Functions (HR, Admin, Finance)				
Key Sources				
Editorial Staff	Reporters/Writers/Journalists	Syndicated Journalism	Web Developers	
Marketing & Sales Staff	Customer Service Staff	Engage BPO for Cust Service	Digital Content Developers	
Printing Press	Printing Press	Outsource Admin Support	Digital Rights Management Splsts	
Support Staff	HR, Finance, Admin Staff		IT/IS Experts	

TABLE 9.1 (Continued)

Business Transformation: From Traditional business to Digital business by Digital Transformation			
Traditional Model	Transition +/- Outsourcing	Digitized Model	Digitalized Model	Digital Transformation
Key Partners				
News Agencies	Media Partners	Social Media Partners	Content Partners	Cloud Services Partners
Printers/Publishers	Printing Partners	Gift Coupon/Vouchers Partners	Ecommerce Partners	Managed Services/BPaaS Partners
Corporate Groups/Business	Corporate Accounts	Syndicate Partners	Payment Gateway partners	BPM Partners
				Analytics Partners
Cost Structure				
Editorial & Production Cost	Production Cost	IT Infrastructure costs	Digital Edition cost	Mobile App Dev./Maint costs
Printing & distribution cost	Printing Cost	Website cost	Digital Distribution costs	Digital Assets/Platform Costs
Distributors + Agent Commission	Distribution Cost		Payment Gateway costs	Bunding Costs
Mgt +Admin Cost	Marketing Cost		Content Provider Costs	DRM Costs
Revenue Streams				
Subscriptions Revenue	Subscriptions revenue		Digital Subscribers Revenue	All Digital Subscribers
Advertisements Revenue	News Stall Sales		Media Partners revenue	Converged Digital Media revenue
	Advertisement Revenue		Reduced Print edition revenue	

Source: Adapted from Amamoo-Otchere, 2016.

of value adding activities that are integrated together in providing the customer with better value at the lowest cost. Here the focus is on creating better value to customers by linking value-adding processes of a firm or network of firms including the partners as a value chain.

In summary, supply chains follow the product from the supply to the customer whereas in value chain customer is the starting point and customer needs determine the processes and activities to creating value by offering the right product or services. Modern business thinking with digital capabilities, focus on industry value chains that typically consists series of processes and activities needed from designing, developing, producing or manufacturing products and services according to their different portfolios.

How can we combine the value chain thinking of a firm with digitization of products and services? One way to look at it is to view the BM of the firm as digitization of market and customer access while the operating model of the firm shall focus on digitization of value chain and processes. Value chain in a digital context can also be seen as vertical and horizontal value chains. This is similar to vertical and horizontal integration of supply chains in the product context. Vertical value chain the firm owns and conducts all the operations and processes needed to design, create, and deliver the product or service end to end. This will enable the business firm efficiently manage the supply and demand and sustain quality with better cost advantages (economies of scope). For example, Apple produces the top mobile with software, content, and all that needed for superior mobile experience from design to App Store and iTunes content to retail outlets.

Horizontal value chain focuses on one or more select operations that can serve different markets and industry with same or similar products and services. Horizontal value chains help firm achieve cost efficiencies by specializing select competencies (economies of scale). It is vital to understand that management strategies and structures for vertical and horizontal value chains are quite different especially in digitized products and services. For example, Amazon has established a complete marketplace for all products and services specializing in distribution of products and services instead of getting into design and production of products and services.

In the product-based business, vertical value chains, and horizontal value chains may result in monopolies and dominant market players respectively. However, in digitized BMs for products and services, the same can be termed as ecosystems and platforms. Digitally connectivity and increased adaption online and internet technologies by businesses have given raise to business networks and highly binding business collaborations (economies of speed).

This in turn is popularizing what is called business ecosystem and platform BMs. How these two are related?

Business ecosystem as a concept appeared first in a Harvard Business Review article by John F. Moore in 1993. As commonly defined in many later articles, business ecosystem is in principle a network of interlinked business entities. These entities may comprise of all the suppliers, distributors, vendors, and other parties who are part of the value chain. All of them actively interact in a complementary way within the boundaries of their BMs and value propositions as related to their products or services.

Platform BM is a recent phenomenon. It is an expanded case of BM that creates value by facilitating exchanges between two or more interdependent groups, usually consumers and producers (Moazed and Johnson, 2016). These platform BMs are very dominant in today's context and more than half of S&P 100 companies with very high market capitalization are based on these technology platforms. Some of the popular technology platforms are Amazon, Apple, Alphabet(Google), Amazon, Alibaba, Facebook, LinkedIn, Netflix, Salesforce Tencent and Twitter, which are obviously very highly valued global business entities today. Of course there are other non-tech platforms also such as Bosch, Disney, GE, Merck, and Schneider to name a few. As per International Data Corporation (IDC) report predictions, more than 50% of large enterprises by 2018 will be create or partner with industry platforms, especially those enterprises with mature digital transformation strategies. Figure 9.3 depicts the relations between and business ecosystems and platform BMs.

As a case of illustration for business ecosystems and platform models, Industry 4.0 can view as a best example. Concept of Industry 4.0 and smart factories are emerging examples of industry platforms. Starting 21st century, information technology (IT) advancements have taken the world by storm. It is expected that by year 2020, industrial world will adapt to concepts of smart factories due to the coming together of computers with machine learning algorithms and robotic automation. These innovations in digital technology and smart connections have led to a collection of platform BMs. This constellation of technology platforms is commonly referred as Industry 4.0. Industry business leaders are contemplating and embracing the newfound strategies for digital supply chains, intelligent products, and services manufactured in smart factories and automation technologies.

PwC report 2016 on new trends in industrial-manufacturing, namely Industry 4.0: Building the digital enterprise, elaborates on the possible use cases and future benefits of digitizing horizontal and vertical value chains across industry verticals that involve business ecosystems and platform BMs (Geissbauer, Vedso, and Schrauf, 2016). This is what actually means by digitizing horizontal and vertical value chains. Internal operations within a company across the value chain-inbound and outbound logistics, operations, marketing, and sales, customer service-need to be integrated and transformed using digital technologies and automated processes to gain operational efficiency, cost reduction, and superior quality for enhanced customer satisfaction. This is called digitizing the horizontal value chain and industries in Japan and Germany are in the forefront of such digital transformation using Industry 4.0.

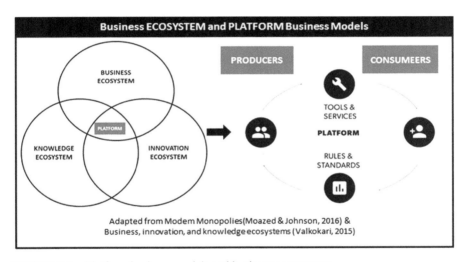

FIGURE 9.3 Platform business models and business ecosystems.

9.5 DIGITAL BUSINESS MODEL (BM) CANVAS FOR BUSINESS INNOVATION

Digital BMs are considered as business propositions by companies that promote solutions and experiences using digital technologies instead of simply attempting to provision of products and services. Revisiting the definitions of digitization and digitalization, Gartner defines digitalization as "the use of digital technologies to change a BM and provide new revenue and value-producing opportunities." Also, digitization of processes and resources through platforms ad ecosystems helps to digitalization the BM (Williams, Chatterjee, and Rossi, 2008; Berman, 2012). Current day BMs area is driven by demanding and well-informed customers in all markets and in every industry verticals such as financial and banking services, healthcare, entertainment, transportation, and government services. Putting all these contexts together, it is now logical to see why businesses are moving their focus to digitized information systems, IT-enabled business processes, digital platforms, and ecosystems.

Importance of these digital platforms as a new BM mainly started with digital and communication network industry such as telecom, media, entertainment products, and services. However, now all other industry verticals are fast embracing digital BMs where innovation is the key differentiator.

While some digital BMs have taken off to a flying start, many digital BMs fail and become inconsistent. Researches and market analysis show there are many reasons for such failures and some reasons are more pronounced:

- Not considering all alternatives and strategic choices that are peculiar to digital BMs;
- Incomplete preparation for value delivery and value capture through the value creation business logic is sound;
- Wrong and flawed assumptions while designing the BM components;
- Over-emphasizing only one component or part of the BM;
- Not understanding the network effects and network dynamics of the business ecosystem comprising various partners and actors.

Innovating digital BMs can boost the success rates of business. This can be done by taking advantage of digital platforms and focusing the attention to innovating and redesign of key components of the BM for better products and services. Digital BMs and continuous innovation is a perfect combination to increase firm capabilities with more easily networkable, adaptable, and sustainable business agility. However, the challenge is how to express digital BMs in simple way with practical significance. Extending the very popular BM Canvas model proposed by Osterwalder and Pigneur (2010) Figure 9.4 provides a complete perspective of digital BM canvas. By adding digital capabilities to each of the BM components, a comprehensive digital BM canvas shall be envisaged.

9.6 CONFLATING DIGITALIZATION AND BUSINESS INNOVATION-543 GRID MODEL

Combining the discussions so far, we then proceed to conflate different dimensions of BMs, business, transformation, digital transformation, and business innovation. For this, we adopt an integrated approach more from an organizational perspective and innovation capabilities dimensions. We also draw influences from entrepreneurship and technology management studies along with contemporary studies on BMs and BM innovation (Morakanyane, Grace, and O'Reilly, 2017; Schoemaker, Heaton, and Teece, 2018; Bashir and Verma, 2019). Reason for the focus on innovation capabilities dimensions is that many contemporary studies focus merely on one of the four dimensions names Value, Innovation, Digital, or Enterprise dimensions. To do so, first, we conceptualized three levels of business innovations ordered hierarchically and combine them as a conflation of two top levels-AMPEE framework with 5 dimensions, VIDE framework with 4 dimensions which is followed by a third order LEO framework with 3 dimensions. That's why we call it as a "543 Grid Model" for Business Innovation as portrayed in Figure 9.5.

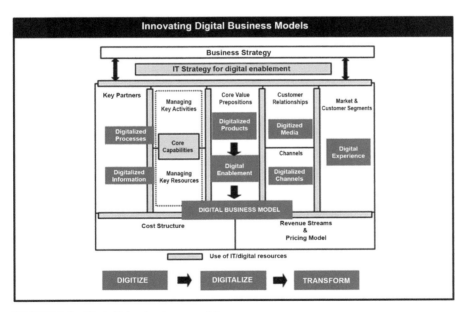

FIGURE 9.4 Digitalizing business model canvas.

This new grid model-543 Grid-with three hierarchical dimensions present an integrated view that can be equally applied to different contexts-Large multinational companies or SMEs, product companies or service providers, traditional organizations or digitally advanced NextGen businesses. By triangulating strategic, tactical, and operational imperatives, it is clear in a hierarchy to understand business innovation and digitalization. This starts from understanding BM as part of AMPEE-Analyze, Model, Process, Execute, Evaluate as five elements of first order. BM as shown is central to understanding business innovation further which is explained by VIDE as a constellation of four elements. This is in turn gives raise imperatives and measures of innovation leadership, innovation enablement and innovation orchestration as three LEO elements. In summary, this 5-4-3 or simply 543 grid model that connects various dimensions of BM and business innovation will definitely a complementary contribution to existing body of knowledge and comprehensions of business leaders towards digitalizing business innovation.

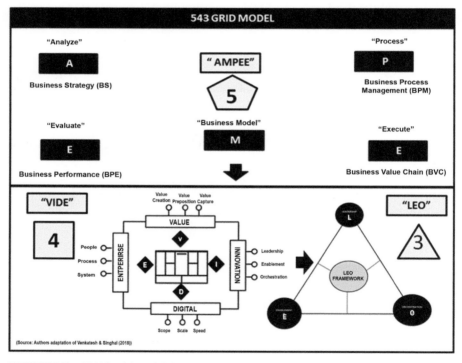

FIGURE 9.5 "543" grid model for business innovation.

9.7 CONCLUSIONS

Conflation of extant knowledge of BS, innovation management, digital transformation, entrepreneurship management, and service BMs are seen as practical approach for closing some of the gaps in contemporary digitalization and business innovation insights. Lack of common integrated innovation frameworks and metrics of innovation capabilities are limiting the use of BM as units of analysis, the fact which is exemplified by recent business research (Lambert and Davidson, 2013; Wirtz, Gotten, and Daiser, 2016). By providing a new perspective of business innovation as a 543 Grid model, this chapter conclusively leads to two distinct output and findings. First, being the fact that the future commentaries and research on digitalizing business innovation will be more effective in a multidimensional approach that involves dimensions of value proposition, innovation capabilities, digitalization imperatives, and enterprise perspectives as part of higher-order AMPEE dimensions and VIDE dimensions. Secondly, measures of digitalization and innovation capabilities now can be studied as per the LEO framework that explains the dimensions of innovation leadership, innovation enablement, and innovation orchestration. This integrated "543 Grid model" approach extending from the conflation of three orders of business innovation and digitalization dimensions will sure lead to practical insights for managing challenges for different approaches to services business such as managed services; IT enabled services, business process outsourcing services, and digitally-enabled business-process-as-a-service models (Venkatesh and Singhal, 2019).

From the practical standpoint, we also recognize that the multi-disciplinary nature of business innovation spanning across size, industry, and innovation priorities are vastly driven by digital capabilities. Digitalizing business innovation, especially by SMEs in emerging markets need to undertake practical steps towards business innovation and digital transformation at business units or instance level. Some of the business innovation and digitalization constructs, as posited by this proposed grid model, can further be fine-grained as suitable to other vertical domains and service sectors with further research contribution. In essence, the proposed "543 Grid Model for Business Innovation" can be used as a good guiding point for the detailed development of operational plans and business process reconfigurations towards digitalizing business innovation and transformation that are possible by getting more digital in all aspects. Salient arguments and propositions of this chapter can also boost a better understanding of digital platform BMs.

Given the newness and explorative nature of this subject, explanations, and the discussions around the proposed grid model for business innovation should be considered as highly suggestive for further extended research and investigation. This chapter also does not cover all aspects of digital innovation constructs. Further empirical studies and business cases will surely have a significant contribution to theory and practice to extend the applicability of this 543 Grid model.

KEYWORDS

- **business innovation**
- **business model**
- **business transformation**
- **digital business models**
- **digital transformation**
- **digitalization**

REFERENCES

Amamoo-Otchere, E. [Emmanuel]. (May 7, 2016.) Posts [Digitalization & Digitization Value Propositions]. Retrieved December 19, 2017 from https://www.linkedin.com/in/eaojnr/

Andersson, J., & Lyckvik, L., (2017). *Business Model Innovation-Challenges and Opportunities in the Swedish Newspaper Industry*.

Ashurst, C., Freer, A., Ekdahl, J., & Gibbons, C., (2012). Exploring IT-enabled innovation: A new paradigm? *International Journal of Information Management, 32*(4), 326–336.

Baden-Fuller, C., & Haefliger, S., (2013). Business models and technological innovation. *Long Range Planning, 46*(6), 419–426.

Bashir, M., & Verma, R., (2019). Internal factors and consequences of business model innovation. *Management Decision, 57*(1), 262–290.

Berman, S. J., (2012). Digital transformation: Opportunities to create new business models. *Strategy and Leadership, 40*(2), 16–24.

Bharadwaj, A., El Sawy, O. A., Pavlou, P. A., & Venkatraman, N. V., (2013). *Digital Business Strategy: Toward a Next Generation of Insights.*.

Bouwman, H., De Reuver, M., & Nikou, S., (2017). *The impact of Digitalization on Business Models: How IT Artefacts, Social Media, and Big Data Force Firms to Innovate Their Business Model*.

Bouwman, H., Nikou, S., Molina-Castillo, F. J., & De Reuver, M., (2018). The impact of digitalization on business models. *Digital Policy, Regulation and Governance, 20*(2), 105–124.

Casadesus-Masanell, R., & Ricart, J. E., (2010). From strategy to business models and onto tactics. *Long Range Planning, 43*(2), 195–215.

Cavalcante, S., Kesting, P., & Ulhøi, J., (2011). Business model dynamics and innovation: (Re) establishing the missing linkages. *Management Decision, 49*(8), 1327–1342.

Clauss, T., (2017). Measuring business model innovation: Conceptualization, scale development, and proof of performance. *R&D Management, 47*(3), 385–403.

Demil, B., & Lecocq, X., (2010). Business model evolution: In search of dynamic consistency. *Long Range Planning, 43*(2), 227–246.

Dobni, C. B., (2008). Measuring innovation culture in organizations: The development of a generalized innovation culture construct using exploratory factor analysis. *European Journal of Innovation Management, 11*(4), 539–559.

Euchner, J., & Ganguly, A., (2014). Business model innovation in practice. *Research-Technology Management, 57*(6), 33–39.

Frankenberger, K., Weiblen, T., Csik, M., & Gassmann, O., (2013). The 4I-framework of business model innovation: A structured view on process phases and challenges. *International Journal of Product Development, 18*(3/4), 249–273.

Geissbauer, R., Vedso, J., & Schrauf, S., (2016). *Industry 4.0: Building the Digital Enterprise*. Retrieved from: PwC Website: https://www.pwc.com/gx/en/industries/industries-4.0/landing-page/industry-4.0-building-your-digital-enterprise-april-2016.Pdf (accessed on 21 October 2020).

Hansen, M. T., & Birkinshaw, J., (2007). The innovation value chain. *Harvard Business Review, 85*(6), 121.

Heikkilä, M., Bouwman, H., & Heikkilä, J., (2018). From strategic goals to business model innovation paths: An exploratory study. *Journal of Small Business and Enterprise Development, 25*(1), 107–128.

Knight, D., Randall, R. M., Muller, A., Välikangas, L., & Merlyn, P., (2005). Metrics for innovation: Guidelines for developing a customized suite of innovation metrics. *Strategy and Leadership*.

Lambert, S., (2015). The importance of classification to business model research. *Journal of Business Models, 3*(1).

Moazed, A., & Johnson, N. L., (2016). *Modern Monopolies: What it Takes to Dominate the 21st Century Economy*. St. Martin's Press.

Morakanyane, R., Grace, A. A., & O'Reilly, P., (2017). *Conceptualizing Digital Transformation in Business Organizations: A Systematic Review of Literature.*

Müller, J. M., Buliga, O., & Voigt, K. I., (2018). Fortune favors the prepared: How SMEs approach business model innovations in Industry 4.0. *Technological Forecasting and Social Change, 132*, 2–17.

Nadeem, A., Abedin, B., Cerpa, N., & Chew, E., (2018). Digital transformation and digital business strategy in electronic commerce-the role of organizational capabilities. *Journal of Theoretical and Applied Electronic Commerce Research, 13*(2), i–viii.

Osterwalder, A., & Pigneur, Y., (2010). *Business Model Generation: A Handbook for Visionaries, Game Changers, and Challengers.*.

Pandya, K. V., & Anand, H., (2008). Role of innovation in IT in achieving its business objectives: A case study. *International Journal of Business Innovation and Research, 2*(3), 289–313.

Schoemaker, P. J., Heaton, S., & Teece, D., (2018). Innovation, dynamic capabilities, and leadership. *California Management Review, 61*(1), 15–42.

Spieth, P., Schneckenberg, D., & Matzler, K., (2016). Exploring the linkage between business model (&) innovation and the strategy of the firm. *R&D Management, 46*(3), 403–413.

Spithoven, A., Vanhaverbeke, W., & Roijakkers, N., (2013). Open innovation practices in SMEs and large enterprises. *Small Business Economics, 41*(3), 537–562. Retrieved from: http://www.jstor.org/stable/43552884Copy (accessed on 21 October 2020).

Trimi, S., & Berbegal-Mirabent, J., (2012). Business model innovation in entrepreneurship. *International Entrepreneurship and Management Journal, 8*(4), 449–465.

Valkokari, K., (2015). Business, innovation, and knowledge ecosystems: How they differ and how to survive and thrive within them. *Technology Innovation Management Review, 5*(8).

Venkatesh, R., & Singhal, T. K., (2019). Conflating the dimensions of business innovation and digital transformation. *Amity Global Business Review, 47*.

Venkatesh, R., Mathew, L., & Singhal, T. K., (2018). Imperatives of business models and digital transformation for digital services providers. *International Journal of Business Data Communications and Networking (IJBDCN), 15*(1). Accepted and Published.

Venkatesh, R., Singhal, T. K., & Mathew, L., (2019). Emergence of digital services innovation as a path to business transformation: Case of communication services providers in GCC region. *IJITEE (International Journal of Information Technology and Electrical Engineering), 8*(6c), 56–63.

Venkatraman, V., (2017). *The Digital Matrix: New Rules for Business Transformation Through Technology.* Greystone Books.

Whitelock, K., (2017). *Monetizing Digital Services and Partner Ecosystems*. Retrieved from: https://www.bearingpoint.com/files/Monetizing_Digital_Services_EN.pdf (accessed on 21 October 2020).

Williams, K., Chatterjee, S., & Rossi, M., (2008). Design of emerging digital services: A taxonomy. *European Journal of Information Systems, 17*(5), 505–517.

Wirtz, B. W., Schilke, O., & Ullrich, S., (2010). Strategic development of business models: Implications of the Web 2.0 for creating value on the internet. *Long Range Planning, 43*(2), 272-290.

Wirtz, B., Göttel, V., & Daiser, P., (2016). Business model innovation: development, concept, and future research directions. *Journal of Business Models, 4*(1).

World Economic Forum, (2017). *Digital Transformation Initiative Telecommunications Industry*. In World Economic Forum.

Zott, C., Amit, R., & Massa, L., (2011). The business model: Recent developments and future research. *Journal of Management, 37*(4), 1019–1042.

CHAPTER 10

Digital Payment: A Robust Face of Modern India

PRIYANKA JINGAR,[1] RAVINDAR MEENA,[1] and SACHIN GUPTA[2]

[1]Research Scholar, Department of Business Administration, Mohanlal Sukhadia University, Udaipur, Rajasthan, India

[2]Assistant Professor, Department of Business Administration, Mohanlal Sukhadia University, Udaipur, Rajasthan, India

ABSTRACT

Our nation has witnessed a remarkable augmentation in digital payments. Mounting government initiatives such as Digital India is playing a crucial role in stepping forward towards digitalization. The government aims to craft a digitally empowered nation that is cashless, paperless, and faceless. Electronic transactions made at the point of sale system to a website e-commerce shopping cart either by using mobile banking or internet banking with the help of Smartphones or debit and credit cards are known as digital payment. Embracing cashless payments has been outstandingly amplified by Prime Minister Mr. Narendra Modi by banning the esteemed value paper money. This decisive move has shown an exceptional enlargement in an online transaction. This chapter aims to demystify the concept of "Digital Payment" in our country. At the end of reading this paper, every reader would have lucidity in their mind about the impact of digital payment to bloom the financial technology in India.

10.1 INTRODUCTION

Every disturbance creates opportunities in an environment, and a kind of disturbance was the declaration of demonetization on 8 November 2016. It has shaped the culture of digital payment in India. It provides a robust platform for various digital wallet businesses to welcome the prospect with open hands and to enlarge their segment of the market. Demonetization has offered fresh vistas for the implementation of a digitally enhanced payment system, as an option to paper money. Digital payment refers to a service that facilitates to take out pecuniary dealings electronically. Cashless payments for monetary transactions can be carried over a digital wallet. India has seen an incredible rise in the digit of digital wallet users. The country is progressively touching towards a charisma of a cashless economy. Digital payment is making our life comfortable by helping to carry out transactions swiftly and effortlessly.

The Reserve Bank of India formed a five-member board reporting Nandan Nilekani committee to kind recommendations to quicken the growth of digital payments in the nation. With the number of digital payment operators expected to grow to 300 million from 100 million, the committee stated that the per capita digital transactions, which mounted at 22 in March 2019, are probable to rise to 220 by March 2022. These figures are an ideal exemplar of the rising online transactions in the countryside that makes our nation a pacesetter in digital payments, globally.

The central aim of digital payment is to persuade cashless and paperless banking across the homeland. The object is to assist in widening the financial inclusion mission by bringing undersized businesses and low-income family unit into the umbrella of financial services. Digital Payment is an exceptional idea of the Government of India with the ambit to reach customers through their cell phones rather than traditional bank branches.

After demonization, the Indian banking sector has undergone digital renovation called payment applications. It means electronic currency or digital wallet. It relied on the design of Pre-paid instruments (PPI) where user can load money into the digital wallet and employ it to execute a range of dealings such as shopping, transferring money, utility, and bill payment, travel, and entertainment services, QR code scan and pay, etc.

10.2 DIGITAL PAYMENT: A RISING TREND

Digital payment procedures are extremely convenient, simple to make and provide a platform for customers and vendors to make payment procedure flexible. It is a widely accepted alternative to long-established way of payment. It facilitates to make payments from anywhere and at anytime. After the demonetization, individuals gradually started accepting online payments via electronic or digital mode. Termination of old currency instigates the need of digital payment in our nation. In nutshell, digital payment takes place when products or services are purchased with the help of diverse electronic mediums. There is no use of currency notes or bank cheques in digital mechanism.

The digitized payment market is a major milestone of technological modernization driven by customary financial establishments. Digital payment market is a conquering system to brand new technologies that endlessly convene with others and increase in the ascend of innovative payment methods and subsequently generating business opportunities. Policymakers around the globe are taking remarkable attention to discover the opportunity of moving towards a cashless and paperless financial system. Digitization of payment is the paramount approach to shift towards a cashless economy.

10.3 NEED FOR DIGITAL PAYMENT APPLICATIONS

The long-established concept of commerce was the right product must be accessible at the correct location and at the accurate moment is now reinstating with an elastic concept that is product or services must be obtainable everywhere and anytime. This vivacious theory has instigated the need for digital payment applications in India. Worldwide, around 1.7 billion adults do not have any bank account at any financial establishment. Our country possesses a large number of unbanked populations that is the second largest in the world, i.e., 19 cores after china 225 million came forward with the need for a digital renovation all over the nation. Around 48% of bank accounts in India are un-operated from the last one year. Ratio of inactive accounts in our nation is the worlds highest that is two times the average of 25% for developing nations.

The digital payment system has emerged as an essential instrument for advancing financial inclusion since it minimizes the expenditure of providing monetary services to deprived people and amplifies the safety.

The intent is to serve the unbanked and underbanked civilians of India. For those people who are unbanked, digital payment applications will provide much-needed admission to banking services on a mobile phone. It is unfeasible for a traditional bank to open branches in every small village. Still, the mobile phone is an effectual and low-cost dais for taking banking services to each rural citizen even though the Pradhan Mantri Jan Dhan Yojana (PMJDY) has brought down a generous amount of unbanked individual's in India, there are still millions who do not have access to formal financial services. India is home to 21% of the world's unbanked adults, and Payment applications are a footstep to redefine banking in India. It aspires to serve a low-income family unit, drifter workers, and tiny businesses to get them into the formal financial system and bring banking to the masses on a wider range. It provides a perk of secured, technology-driven transactions which can easily be tracked anywhere with no dodge for black money and scam (Figure 10.1).

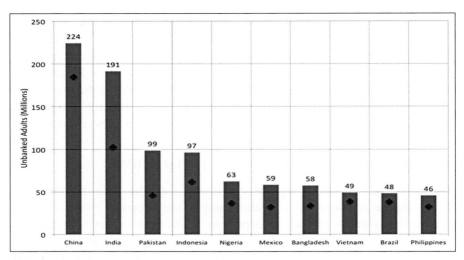

FIGURE 10.1 Number of unbanked adults globally.
Note: Diamonds point towards those unbanked adults with mobile phones.
Source: Global Findex Report, World Bank (2017), from Kunt, 2018.

The above picture clearly shows the need for a digital payment system in India. First time in Indian history, RBI has approved banking authority to another non-banking financial sector. This plan is intended at redefining the Indian economy by providing a protected payment gateway for all transactions. A digital payment reaches out to the nomad laborers and lower-income groups by providing all services on mobile. After demonetization, this is the RBI's initial shift to get rid of black money and persuade cashless transactions to digitize India. Since it is an inventiveness to go cashless, it eliminates the requirement to swap currency notes and the cost of regeneration of unmarked notes.

Change is the only thing permanent and the evolution to digital dealings and development is likely to propel India into the club of developed nations through a range of benefits:

- **Economic Development:** Expedient cashless transactions facilitate individuals to make additional purchases in a short span of

time. Countless people end up purchasing items impulsively, thus digital payment is contributing to economic development. Mounting demand for goods will lead to further production and supply which in turn will create jobs. Supplementary digital transactions also imply that additional data will be available to the government for research and enhanced framing of policies. It could be a massive game changer in our economy. It facilitates to check the corruption and the whole financial growth can be traced in a designed way.
- **Reduced expenses:** With increased digital payments, the cost of regenerating and distributing the fresh currency notes can be minimized because the life span of a currency note is around six years and printing of both coins and notes get expensive if the digital payment system does not take over. There is also a vast expenditure involved in labor-intensive accounting of cash at different levels, which can be done away with the cashless economy. The danger of burglary can also be reduced if individuals hold a lesser amount of cash.
- **Transparency and handiness:** Digital payments guarantee answerability in all the dealings. Since the whole thing is recorded digitally, the Government can verify and track all transactions digitally. For example, when the government made digital smart cards obligatory in case of pension payments replacing the system of physical cash payouts in rural areas, a 47% reduction was reported in corruption.

10.4 DIGITAL PAYMENT MODE

The below-mentioned mode of online payment are presently being promoted across the nation.
- Banking cards;
- Unstructured supplementary service data (USSD);
- Aadhaar enabled payment system (AEPS);
- Mobile banking;
- Unified payments Interface (UPI);
- Internet banking;
- Point of sale (POS);
- Bank prepaid cards;
- Micro ATMs.

1. **Banking Cards/Debit Card:** In India, it is the most widespread means of digital payment system. It may be Visa, MasterCard, RuPay, etc. It offers bi-level verification for protected transactions such as one-time password, etc. Banking cards give individuals the authority to buy things physically, online, through mail and on mobile phones.
2. **USSD:** *99# facility has been commenced to capture the financial services by every mango people across the countryside. User can dial a "universal number (i.e., *99#)" on their mobile screen and manage with the help of a list presented on the screen.
3. **AEPS:** With the help of the Aadhaar Enabled Payment System, a user can move money between two Aadhaar linked bank accounts or withdraw money if the user's bank account linked with an Aadhaar card. Biometric authentication is essential to complete the payment process via AEPS. The vendor also can utilize this mode to accept payment by fingerprint authentication of consumers. The vendor does not require giving any transaction charges if the amount is below Rs. 2000.
4. **Internet Banking:** It is an online payment system, facilitates users to carry out a variety of monetary dealings through the financial institution's website. It is further classified into:
   - National electronic fund transfer;
   - Real time gross settlement;
   - Immediate payment service;
   - Electronic clearing system.
5. **UPI:** It is a mechanism that handles several bank accounts bent on a sole mobile application. It is launched by NPCI in India. UPI is one of the most groundbreaking digital payments.

    The above diagram shows year-wise growth of UPI transaction by volume (As on March 2019). As we can see, there is a tremendous growth of UPI transaction volume from 2016 to 2019. It clearly depicts the mounting use of digital payment mechanism.
6. **Digital Mobile Wallet:** An intelligent approach to transport paper money in a digital system. Facilitates users to feed and connect their card information to wallet applications. It reduces the significance of plastic cards as users can make payments by means of an online digital wallet. Many companies have their digital wallet such as Paytm, Amazon Pay, Airtel Money, M-Pesa, Jio Money, Freecharge,

Mobikwik, Oxigen wallet, etc. Paytm was the first company in India to start a trend of mobile wallet.

7. **Bank Pre-Paid Cards:** It is known as Gift Card or Pre-loadable debit card for a single-use or re-loadable for numerous uses. It is primarily used for corporate gift, reward card or any single-use card for gifting purposes.
8. **POS:** It is the place where a payment took place. There are two types of POS machine available, one is wired and another one is wireless. It may be a fuel station, shopping mall, kirana store, movie booking counter, etc.
9. **Mobile Banking:** A service offered by a bank to facilitate its clientele to perform numerous kinds of pecuniary dealings by means of a cell phone. Every bank equips its users with Android-based, Windows-based, and Apple store ready mobile applications for mobile banking.
10. **Micro ATMs:** It enables instant transaction of payment. Micro ATMs permit an individual to immediately deposit or withdraw money regardless of the bank linked by a business correspondent.

10.5 MOBILE PAYMENT APPLICATIONS: ACCELERATING INDIA'S DIGITAL GROWTH

The cashless economy of India is duly recorded in November 2016 when demonetization occurs. Digital payments and e-wallet is a blooming landscape in accelerating digital growth in our nation. Our country is the second global leader in the mobile phone market after the United States of America. Below mentioned are few mobile payment applications that are lending a hand to make India a digitally empowered nation.

10.5.1 PAYTM: NEW FACE OF DIGITAL ECONOMY

It is an Indian e-commerce payment mode company. Paytm is extensively accessible in 10 different Indian languages. Paytm offers a pool of services such as Paytm Mall, mobile recharges, entertainment service, movies, in-store payments at grocery outlets, gold saving plan, amusement park tickets, hotels, and restaurants, donation, devotion, fruits, and vegetable shops, pharmacies, and educational institutions with the Paytm QR code, etc. In August 2015, RBI gave a permit to commence a Paytm Payments bank. Over 7 million vendors across India are using QR scan to acknowledge payment directly into their linked bank account. Paytm enjoyed an unbelievable enlargement after the termination of huge value currency notes, i.e., 500 and 1000. The lack of cash in hand and extensive queue has made digital wallets an essential requirement for an ordinary people at least those with mobile phones. Relocate money using Paytm is very convenient and hassle-free for the user. A user can track balance, load money to the wallet, transfer money to friends and family, and accept money securely on Paytm. Paytm function in two diverse ways. First is Paytm digital wallet and another one is Paytm payments bank. Paytm wallet is a digital payment tool which permits user to transmit currency through debit card, credit card, online banking. Wallet smooth's the process to pay all bills without giving cash anywhere. For this, the user needs to transmit money from a linked bank account to a digital wallet. Users can load 10,000 rupees in a month, which is extendable up to 1 lakh after fulfilling KYC norms. Another one is the Paytm Payments bank, which acts like any other bank. Money transfer on Paytm is powered by UPI, an innovative tool which allows payment to bank account directly using the mobile number.

The above image clearly states that Paytm has the highest wallet user in India as compared to other payment applications.

10.5.1.1 PROMOTIONAL STRATEGIES OF PAYTM

Paytm refers to Pay through Mobile. Paytm has targeted all individuals irrespective of age, gender, income, or status because it wants to make a way into every alcove of India. It started its operations as a B2B business but later on realized the importance of customer involvement and opened B2C option. These are a few promotional strategies adopted by Paytm to reach masses:

- #PaytmKaro is getting viral nowadays and it is united with NDTV and Uber for demonstration. By going live on Facebook and creating awareness camp, people are getting knowledge regarding digital wallets.
- Paytm is the first company to commence a fashion of cashback. Instead of giving a discount, they offer money-back in the digital wallet.
- Paytm has adopted an innovative and unique promotion policy to construct better brand visibility. They launched several schemes and initiated net banking and offers as an ingredient of their promotions. The advertisement was shown via numerous channels on radio,

television, newspapers, magazines, and billboards. Mouth publicity also plays a most important role in promotion.
- Paytm collaborate with Mumbai dabbawallas, enabling people to use the app, Paytm cash alongside lunch service.
- Celebrity endorsements are done by Paytm to grab new user attention.
- Wallet of Paytm can be used on the IRCTC website.
- Demonetization of currency notes worked optimistically in its support and garnered massive publicity.

The above image shows the number of card transactions via Paytm v/s Indian banks. Paytm users are gradually increasing from 2016 to 2018.

10.5.2 BHIM: MAKING INDIA CASHLESS

BHIM (Bharat interface for money). BHIM is using UPI to perform the transaction in an effortless, simple, and speedy manner. It is a mobile application through which a user can transfer money to any person on UPI by means of scanning QR code and UPI ID. Users can also request for money from a UPI ID. BHIM app is developed by the National Payments Corporation of India. This money transfer application is launched by our Prime Minister on 30 December 2016 in New Delhi at Digi Dhan Mela to fetch financial inclusion to the country and to make people digitally empowered. It is intended to soften the progress of online payments. Money transfer on BHIM app is instant and can be accessible 24*7.

10.5.2.1 PROMOTIONAL STRATEGIES OF BHIM

The Government of India has launched the following schemes to reward BHIM users:
- Digi-DhanMela as a promotional tool was used by Hon'ble Prime Minister Narendra Modi and also asked Smartphone manufacturer to prior install the BHIM application in mobile phones.
- BHIM Cashback Scheme for Individuals.
- An incentive for on boarding over BHIM App will be valid for BHIM App users only. The motivation will be rewarded to the fresh users of the BHIM app who effectively download the BHIM app, link the BHIM app with their bank accounts, and originate ten successful unique monetary transactions through the BHIM app amount of Rs

50 or more. Cashback will be paid to eligible only for the new BHIM app users, i.e., those who download the application for the first time and make the particular quantity of transactions between the precise periods beginning from 4 July 2018 till 31 March 2019.

10.5.3 GOOGLE PAY: MONEY MADE SIMPLE

It is developed by Google. It assists the user in crafting payments with Smartphones, tablets, or watches. Google Pay initially launched on 11 September 2015, in the name of Android Pay but later folded into the new Google Pay app on 28 August 2018. Google launched a mobile app payment service to target the users in India called Tez. Tez functioned on an Android app supporting English, Hindi, and other different languages of India. As per the data of October 2017, more than 30 million transactions were recorded on the application, and recently, monthly active users on Google Pay reached a whopping number of 65 million users and recorded $100 billion in digital transactions in India. Google Pay is paving the digital growth of our nation.

10.5.4 PHONEPE

PhonePe was launched in 2015. PhonePe is an electronic payment system and a digital wallet app. Its headquarters is situated in Bangalore, India, and it is the primary money transfer application built on UPI. In 2016, Flipkart acquired this company. It is widely available in 11 different tongues. PhonePe lets consumers transmit and accept money and offer a wide range of services such as recharge, bill payment, etc.

10.5.5 MOBIKWIK

MobiKwik was founded in 2009. It is an Indian digital wallet company that provides a mobile phone-based payment system. It's headquarter is situated in Gurugram. MobiKwik is India's second-largest digital wallet player, and it is also considered along with the apex 3 players in the huge payment gateway business.

The MobiKwik wallet can be used for bill payment, recharges, taxi booking, Bus/ train/ flight ticket booking, grocery payment, payment to online and offline retail stores. Its purpose is to bridge the digital gap among customers and companies by providing bill payment at no extra cost. MobiKwik is the best way to pay digitally, whether it is life insurance, prepaid, post-paid, data card, gas, water, broadband, electricity, and so on. It assists in one touch money transfer and facilitates its users to store their hard-earned money.

10.5.6 AMAZON PAY

The worldwide colossal, i.e., Amazon has launched its own digital wallet recognized as Amazon Pay. Globally, Amazon Pay was launched in 2007, and after a span of 10 years, it entered into the Indian market a decade later, i.e., in 2017. It is owned by Amazon to facilitate users to pay online. The headquarters of Amazon is situated in Seattle, Washington, United States. Amazon Pay's digital wallet helps to make payments for diverse dealings. It offers users to pay with their respective Amazon account on outside commercial websites. The coverage of Amazon is massive. It is available in 18 different countries, including India. Amazon Pay offers a multiplicity of goods for buyers and businesses to persuade online payment. Amazon Pay offers an opportunity to avail of products and services on the mobile platform as well as on the website.

10.6 STRENGTHS OF DIGITAL PAYMENT SYSTEM

- **Simple and Handy:** Digital payments are simple and handy. There is no need to carry currency notes. Digital payment applications play an alternative to cash. UPI applications and online wallets are making digital dealings trouble-free.
- **Transact or Pay Money from Anywhere:** A digital payment application offers 24*7 payment solutions. One can pay, send, or receive money anytime and at anyplace.
- **Security:** The digital payment system facilitates consumers to pay via mobile devices. Consumers' are no longer compulsory to take cash and no longer have to presume the security threat related to cash or worry whether they have sufficient cash in their bulky physical wallet. Digital payments reduce the theft risk of having cash in hand. Moreover, mobile payment is a secure means to pay. Credit card information is not stored on the Smartphone directly

but in the cloud, so no thief could dig out credit card details just by stealing the device.
- **Time-Efficient:** Wire transfers can be tedious or postal may need days to process, whereas online money transfer among different accounts can be done in a minute's time.
- **Written Record:** Recording of cash expenditure takes a large amount of time, but in the case of digital payment, we do not require to maintain our spending.

10.7 DRAWBACK OF DIGITAL PAYMENTS

- **Tricky:** Digital transactions are based on web connectivity, Smartphones, and banking cards as they are a bit tricky for an ordinary people to learn and understand, such as farmers, workers, etc.
- **Restriction:** Each digital payment application has its limitation regarding the maximum transaction per day, as well as the maximum amount in the accounts. Paytm has a limit of loading a maximum of 1 lakh rupee at the day end, and BHIM has a transaction limit of 40,000 Rs every day.
- **The Risk of Being Hacked:** Hackers can easily hack the data associated with digital payment and digital wallet. Hackers can hack our personal information and can access our bank account also.
- **Scarcity of Secrecy:** Every bit of information regarding the amount, time, recipient, etc., is stored in a payment system database. Which means the payment agency can use this information.
- **Mandatory Internet Access:** Without adequate internet connectivity, a user cannot perform digital payments. Proper internet access is an essential requirement for digital transactions.
- **The Difficulty of Transmitting Currency between Diverse Payment Mechanisms:** generally, most of the digital payment systems do not synchronize with one another.
- **Overspending:** Digital payment encourages people to overspend their money.

10.8 CONCLUSION

The digital payment mechanism is changing the face of the digital economy. Digital Payment is ensuring speedy development and expansion of the Indian economy. It is the Future of India's economy. Education and training of digital technology in urban as well as the rural area should be provided in order to use digital payment in routine life. Digital payment is a significant system in our nation promoting digital India started by our honorable Prime Minister Narendra Modi Ji, which increases lucidity of cash. Digital payment is an imperative instrument for encouraging the financial formation and financial inclusion in India. It reduces the expenditure of rendering pecuniary services to an ordinary public also; it perks up the safety and accessibility. It is helping customers as well as vendors to transact online. India is gradually stepping towards a cashless economy.

KEYWORDS

- **Bharat interface for money**
- **cashless transaction**
- **demonetization**
- **digital India**
- **digital payment**
- **pre-paid instruments**

REFERENCES

Anand, N., (2018). *Nearly Half of Indian Bank Accounts are Rarely Used.* Quartz India.
Ash, P., (2017). *Quora.* https://www.quora.com/What-exactly-paytm-is-how-does-it-work-and-how-can-one-use-it (accessed on 21 October 2020).
Bhasin, H., (2018). *Marketing Mix of Paytm-Paytm Marketing Mix.* Retrieved from: MARKETING91: https://www.marketing91.com/marketing-mix-paytm/ (accessed on 21 October 2020).
BHIM UPI. https://www.bhimupi.org.in/who-we-are (accessed on 21 October 2020).
BHIM, (n.d.). Retrieved from: Wikipedia: https://en.wikipedia.org/wiki/BHIM (accessed on 21 October 2020).
Gupta, K., (2018). *New RBI Norms Put Mobile Wallets on Par with Payments Banks.* Live mint: https://www.livemint.com/Industry/KJjmn5NoAczkYvKrL6wpGl/New-RBI-norms-put-mobile-wallets-on-par-with-payments-banks.html (accessed on 21 October 2020).
Kunt, D. (2018). Banking the Masses: 2018 Edition. Money & Banking.
Manohar, K., (2017). *Marketing Strategy Paytm.* Slide Share.

Nidhi, S., (2017). *Four Reasons Why BHIM App is Better Than Other Private Mobile Wallets.* Retrieved from: Entrepreneur India: https://www.entrepreneur.com/article/287895 (accessed on 21 October 2020).

Paytm. https://paytm.com/offer/8756-2/ (accessed on 21 October 2020).

Reserve Bank of India, (2016). Retrieved from: Reserve Bank of India. https://m.rbi.org.in/Scripts/FAQView.aspx?Id=73 (accessed on 21 October 2020).

Sharma, R., (2016). *Gadgets360.* Retrieved from: gadgets 360 An NDTV venture. https://gadgets.ndtv.com/apps/features/what-is-paytm-and-how-to-use-paytm-wallet-1625271 (accessed on 21 October 2020).

Surbhi G., & Garg, M. N., (n.d.). Changing landscape of banking system in India: Payment banks opportunities or challenges. *ELK Asia Pacific Journals.*

The Economic Times, (n.d.). Retrieved from: The Economic Times: https://economictimes.indiatimes.com/definition/payments-banks (accessed on 21 October 2020).

CHAPTER 11

The Emerging Smart Supply Solutions in Fresh Fruits: India Matching the International Business Standards, New Formats, and New Technologies

NAVITA MAHAJAN[1] and FRITS POPMA[2]

[1]*Associate Professor, Amity International Business School, Amity University, Noida, Uttar Pradesh, India*

[2]*Managing Director, Popma Fruits Expertise, The Netherlands*

ABSTRACT

India's Horticulture, which hit a record high of 300 million MT in 2016–2017, which was even more than the production of food grains for the year. The post-harvest losses up to the extent of 40%, coupled with India's position in International Business of Fruit Export, has attracted modern and smart digitized formats in Supply Chain Network technologies. Artificial intelligence (AI) has further scaled the prospective chances to expanded margins through quality regulators. The input channel, business model (BM), re-engineered the Horticulture Industry of our country and losses to the extent of Rs 400 billion a year, gearing up new technologies in supply chain and smart packaging. Not only the well-established organizations but also startups have entered into this Smart Technology integration as an opportunity in fresh fruits is higher than the processed foods. We are matching the standards where India will not be the world leader in the production of fruits but a global leader as well in export and treading of fruits in the International Business Platform.

11.1 INTRODUCTION

Fruits and Vegetables comprise around 90% of India's horticulture production. India stands second in fruit production in the world with production of 300 million Metric Tons (2016–2017) and a world leader in many fruit crops like Mango, Banana, Papaya (Table 11.1).

TABLE 11.1 Fruits and Vegetable Production in India

	Crop	Final Production 2016-2017	3rd Estimates for 2017-2018
Fruits	Apple	2,265	2,371
	Citrus	11,419	12,510
	Mango	19,506	21,253
	Banana	30,477	31,083
	Total Fruits	**92,918**	**97,055**
Vegetables	Onion	22,427	22,071
	Potato	48,605	48,529
	Tomato	20,708	19,377
	Brinjal	12.51	12.83
	Total Vegetables	**1,78,172**	**1,79,692**
Total Horticultural			

Source: https://www.thehindubusinessline.com/economy/agri-business/indias-horticulture-yield-projected-to-rise-2-to-307-million-tonnes-in-the-year-to-june/article24811689.ece.

Amongst fruits, the country ranks first in the production of Bananas (26%), Papayas (44%), and Mangoes (including mangosteens and guavas) (41%).

11.2 LEADING FRUIT PRODUCERS OF WORLD

Fresh fruits and vegetable market in India is no way behind any other product where penetration of foreign brands has happened to such a great extent that Indian consumers have now developed a taste to some of the unique fruit flavors of foreign origin, e.g., US Apple, New Zealand Kiwi, South Africa Oranges, South Asia's fruits like Dragon Fruit, Thai Guavas, Longans, Grapes, Passion Fruits, etc. (Table 11.2).

TABLE 11.2 World Fruit Production: Production-Million MT

Country	Fruit Production
China	155.37
India	82.26
Brazil	37.26
USA	26.36
Spain	17.6
Mexico	17.5
Italy	16.3
Indonesia	16.0

Source: Times of India, Article (2016).

Primarily what we eat today in India, right from our dietary staples to fruits, 30–40% on our land comes from foreign origin only. In terms of Calories, around 30% of our Calorie requirement is met from these brands only.

India's share in world fruit production has been estimated as follows, as per National Horticulture Board. China is the world leader sharing 20% of the world's fruit production, followed by India (12%) and Brazil (7%) (Figure 11.1).

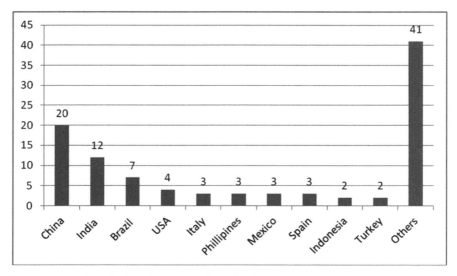

FIGURE 11.1 India's share in world fruit production.

Source: Country-wise production share in percentage terms (compiled from National Horticulture Board Data).

11.3 DRIVERS OF SPOILAGE IN SUPPLY CHAIN OF FRESH FRUITS

It has been estimated that nearly 30–40% of Fresh Fruits and Vegetables Valuing 7000 crores is wasted in India due to lack of Post-Harvest Technologies, lack of storage technologies, very meager level of organized distribution channel network, inadequate cold chain infrastructure. The situation thereof not only presents a challenging situation for the Indian Fruit Industry but an excellent opportunity to set up infrastructure technologies, supply chain networks, and an organized way of handling to minimize post-harvest losses.

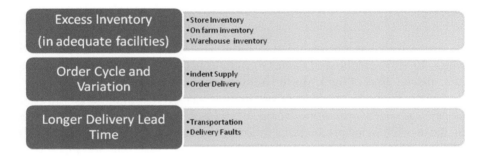

11.4 INDIA'S EXPORT OF FRESH FRUITS

According to statistics released by APEDA, Ministry of Commerce, and India, due to the large product volume of India, there are fabulous prospects for export. During 2018–2019, India exported fruits and vegetables worth 1,470 USD Millions, which comprised of fruits worth 692 USD Millions and vegetables worth 777 USD Millions (Table 11.3).

The major fruits exported from the country are grapes, pomegranates, mangoes, bananas, and oranges. The major destinations are our neighboring countries, the Middle East, South East Asia, and parts of Europe.

India's stake in the worldwide market is still nearly 1–1.5% only; however, there is mounting recognition of horticulture produce from the country. The state of art technology and infrastructure are the key reasons behind this.

The world that demands the fresh fruit Industry to grow and gain efficiency, it is a question of what digital innovation and revolution can be brought into the industry as the growing and digital population in the country cannot afford losses but only increased production.

TABLE 11.3 Major Fruit Export from India-Qty.-000, MT, Val.-Rs. 000, Lakhs

	2016–2017		2017–2018		2018–2019	
	Qty.	Value	Qty.	Value	Qty.	Value
Grapes	198.4713	178	1.88	1.90	2.46	23
Pomegranates	50	49	47.00	54.00	67.00	69.00
Banana	110	38	1.01	35.00	1.34	41.00
Mangoes	52	44	49.00	39.00	46.00	41.00
Oranges	48	11	15.00	4.00	43.00	25.00
Other Fresh Fruits	38.00	11	22.00	6.00	15.00	9.00
Apples	22	6	12.00	4.00	167.00	8.00
Watermelon	26.00	6	26.00	6.00	33.00	7.00

Source: Compiled and calculated from DGCIS annual exports.

The digitized transformation in the fresh fruits Industry can be a success story only with special emphasis on greenhouse technologies, polyhouses, polytunnel technologies, biotechnologies, and modernized logistics.

In a recent event by a major corporate in India, It's the digital bloom that is leading the Indian ecosystem a digital flavor. The rising mobile networks, broadband-based internets, cloud-oriented platforms, information technologies, artificial intelligence (AI), and open data are enabling agriculture and related sectors a transformation, what is required in India's oldest and gadgets adopted profession.

Today, our Agriculture Extension is merging with technology as the IT services sector is helping open up new opportunities for the upcoming entrepreneurs. The technology-based Startups are playing the key drivers in reinventing agriculture, which accounts for half of India's workforce, but only about 14% of its GDP.

11.5 ENTREPRENEURIAL TRENDS

This is an amazing fact that the numbers of agriculture based-entrepreneurs in the country are rising, and many do not have a background in agriculture. The agro-based technology competitions, awards, recognitions, and investments are also emerging in the sector, and Indian youth is being attracted to new technologies, diversified mechanisms of agriculture techniques, alternate cropping patterns, scientific models of the supply chain, and so on. Nevertheless, the challenges seem to be arduous, but need to be acknowledged and tackled.

In spite of the penetration of technology in Indian agriculture, productivity and quality are still the key issues for Indian startups. If we compare with global standards, the productivity of Indian fruits is far behind, due to varietal constraints, old hybrids, soil factors, rain-fed agriculture, application of agronomical practices, post-harvest mechanisms, and poor refrigeration retails supply chain patterns, and on store handling facilities small farm sizes and lack of fairness in financial stakeholders are other challenges.

One of the key recommendations of expert committees on enhancing farmers' income and entrepreneurship has been liberalization of the agriculture value chain, especially post-harvest marketing networks.

11.6 ARTIFICIAL INTELLIGENCE (AI) AND FRUITS SUPPLY MODELS

Using AI the modern retail companies of Fresh Fruit and Vegetable chain such as Ninjacart, Grofers, Crofarm, krishiHub are able to predict demand from the data they have collected, allowing them to minimize waste which may otherwise crop up due to excess stocking, transportation, moisture loss in warehouse or any sort of post-harvest handling. There is no compulsion on farmers to sell their produce to these companies. Payments are made directly into farmers' accounts; hence corruption is no place here (Table 11.4).

TABLE 11.4 Uniqueness of Some of the Modern Supply Chain Models

Company	Unique Features
Ninjacart	• Farmer on Board.
	• Once crop is ready, company offers to procure usually at price around 15% higher than market.
	• Every worker in warehouse has smartphone to see instructions.
Crofarm	• Network with around 6000 farmers.
	• Daily SMS on the quantity of produce required and price offered.
	• Once crop reaches the center, SMS delivered to farmers on arrival confirmation.
	• Direct payment transferred to farmers' accounts.
KrishiHub	• Three apps.
	• One for farmers for maintaining inventory.
	• Another to see orders placed, track payments.
	• Thirds for retailers to place orders and delivery partners to track logistics.

11.7 SMART TECHNOLOGIES INVASION IN FRESH FRUIT INDUSTRY IN INDIA

The amount of wastage in food and loss is estimated to be $ 940 billion and for a country like India the shrinkage in retail supply chain has direct supply chain impact on company's triple bottom line. It is quite possible to reduce the wastages but making this requires a great level of efforts in technology and new systems of handling the fresh fruits so that quality is restored.

Recently many smart organization make up to the extent of 30% have started digitalizing the supply chain network and digitization of supply chain has become the topmost priority.

Today every organization is talking about the importance and positioning of digitalized mechanisms in the fresh supply chain, but in reality only very few have implemented it and turned it into a reality. Today around 75% of organizations are of the opinion that supply chain modernization is driven by cost savings coupled with revenues and supporting new BMs.

11.8 BANANA RIPENING IN INDIA: FROM THE EXPERTS' EYE VIEW

Banana is the second most important fruit crop in India after Mango. It is one of the fruits apart from apple which is demanded throughout the year in India, and the demand goes peak high during Indian festivals. India is the world leader in Banana Production, producing around 15 million MT Banana annually, followed by Brazil, Ecuador, China, the Philippines, etc.

The Indian retail, with a share of around 5%, is a small player in the Indian banana market; relatively speaking, because some 700 tons of bananas are handled per week. "The largest volumes are sold by street stalls; bananas in India remain an important source of nutrients for the poor."

In spite of world leader in production, the export volumes of the country are less. The real problem lies in the variety of this fruit apart the major reason is attributed to ripening technology which was being adopted 5–6 years back. Looking at the problem, Indian Private giants, fruit retail organized supply chain players, multinationals who were in Banana trade have taken the expertise of world level renowned experts, and a major change in Banana ripening has been achieved in the Indian market.

Bananas, during ripening, convert starch into sugars and make the fruit edible. This process releases a lot of heat. The "heat production" or "heat explosion" the temperature development during this stage is high. A single banana can increase with 6–8°C, besides the heat; the banana produces internal ethylene, CO_2, and humidity. This process needs to be controlled.

On a large scale for farmers, it was not easy to afford the technology due to the low price being fetched by fruit but a high price for technology, as per the experience of ripening experts in the country.

The Banana ripening experts have witnessed modern projects in New Delhi with empty banana ripening rooms. While India needs many of these types of rooms. The reason has been essential components were not installed; thus, the banana ripening rooms were not working in the market.

First of all shareholders, from growers to retailers have to understand the banana supply chain. The experts' experience says that in bananas there are two critical points.

- Banana ripening is a race against time; and
- Bananas tell you what to do not the other way around.

Knowing the "yellow" quality of bananas in India, it has been suggested that the first big step to improve was training the farmers, ripeners, and retailers. They should meet in the middle.

Part 2 is to investigate what is available in the domestic market in terms of cooling technology. For experts, it was not necessary to buy a banana cooler outside India, the same with compressors and condensers.

Part 3 is the control software. There is a gap in experience of the Indian software industry but convinced with the right knowledge, the industry will develop a very fast control system. However, the Dutch Industry has more than 50 years' experience in development of ripening software. The Indian banana ripening industry needs another kind of approach. Working with universities and developers of software and experienced advisors in key. The start of the supply chain is different, hot areas and humid areas, 500 to 1000 km away from the big cities, etc.

Last but not least is training ripeners to understand the background of banana ripening, the quality of incoming bananas plus the flow when the bananas left the ripening facility. To increase sales and reduces waste the ripeners need to convince buyers a good ripened banana has a better shelf life, better taste and a better appearance.

Bananas start to ripen themselves after hours when the bunch is cut from the mother plant. The cool chain is key, when the bananas are not cooled back from field heat to 13.7°C, it will create problems as shelf life and taste.

The best option to reduce waste after-ripening is to deliver a good ripened banana which has not much starch inside. This also means our customers; small shop or retailer needs to be convinced yellow bananas sell better.

11.9 SENSOR TECHNOLOGY

This is an emerging technology to detect freshness of fruits. Sensors and IoT in greenhouses are used to monitor the critical parameters of crop growth.

The technology is also used to determine the fruit ripeness and for predicting shelf life of fruits and vegetables during transportation.

The development in sensor technology has led to creation of opportunities for fruits and vegetables producers to find out solutions for various issues such as food safety and food monitoring applications. Some of the remarkable achievements of the technology are reduction in wastage, optimization of produce, reduction in cost and adoption of small devices are encouraging improvements in the food industry.

The ion concentration is measured and that is the principle of measurement of freshness of the fruits.

As the ion concentration changes, we can measure the change in freshness over a period of time.

The method does not require any chemical treatment to fruit for its measurement but can be used in any retail store as well as for frozen products also.

The right stage of ripening determines the usage of the product like as fresh usage, in processing, juices, storage, freezing, drying, etc.

Sensors are playing vital role in food monitoring, apart from remotely monitoring it is essential for the detection of chemicals for determining the determining the freshness of fruit and vegetables (Figure 11.2).

FIGURE 11.2 Sensors to detect fruits and vegetables freshness.

The working mechanism of the sensor mainly governed by circuits and created a software program. There are four different circuits used in the system which are the processor supply circuit, liquid crystal display circuits, the measurement circuit. These circuits are designed by using a circuit design program called Dip Trace after the test is done manually by using circuit test boards. Especially, the measurement circuit is designed by using sensitive and low tolerated resistors in order to increase the sensitivity of performed measurements.

This novel device could be considered as an innovative project which could be a solution for microorganism-caused respiratory diseases and allergy-like diseases by detecting contamination before microorganism colonies become visible with the naked eye by measuring ion concentration changes occurred at samples. Besides, it could be an answer for economic problems of import and export branch, also storage problems of food-related companies, markets, and farmers.

11.10   CA STORAGE AND SUPPLY CHAIN: CASE ON FRESH AND HEALTHY ENTERPRISE, CONCOR INDIA LIMITED, MINISTRY OF RAILWAYS, INDIA

Controlled atmosphere (CA) technology is one of the most important technologies for retaining freshness of fruits. Fresh and Healthy Enterprise Limited, Subsidiary of Concor India, Ministry of Railways-Government of India, has made a remarkable and unique presence in Logistics and Controlled Atmospheric technology, which is based on controlling the ripening mechanism of the fruit by controlling concentrations of Oxygen, Carbon Dioxide and Nitrogen that basically trigger the ripening of fruit naturally. The fundamental deterioration process of fruits is the basic concept behind the development of this technology. As all living produce consume oxygen and produce carbon dioxide, controlling the later is the key feature of this technology.

CA storage is one of the widely used technologies in the International platform and in many countries, on form CA stores are also there in India also, and this initiative has been taken by some of the private players lie Adani Agrofresh, ITC, and Unifrutti India.

- ➢ **Technological aspects of CA:**
  - Construction of storage chambers;
  - Refrigeration equipment for cooling the produce;
  - Maintenance of high RH coolers;
  - Nitrogen generation;
  - Carbon dioxide scrubbers;

- Monitoring and controlling equipment;
- Safety arrangement.

11.10.1 BENEFITS OF CA TECHNOLOGY

Fresh and Healthy Enterprise Limited, was the first organization controlled by the Government of India, that re-engineered the entire Apple supply chain by applying new scientific methods right from procurement to handling, on farm storage, transportation in refrigerated containers hence, adding value to the supply chain of Apple. To facilitate all this, FHEL was installed CA technology in August 2007 in a place called Rai, Kundli, Haryana, India. The strategic location of this CA plant was kept in such a way that it was not only near to Azadpur Mandi, Delhi (only 25 Km), which is Asia's largest fruit and vegetables market, but also that Rai on the Delhi-Haryana border and anytime buyers sellers can jointly facilitate PAN India supply of Apple from this place (Figures 11.3 and 11.4).

FIGURE 11.3 CA Store view from inside.

The main focus of FHEL was on Kinnour variety of Apple, which is found in high regions of Himachal Pradesh. The rationale behind this was that this variety of Apple has a dark red color with high shelf life and unique flavor. This completed Washington Variety of USA Apple and fetched premium prices at every end of the market. Not only for retail/bulk sales, but this variety has also been very much popular among consumers for gifting and corporate gift pack purposes. Apart from this, FHEL also procures other varieties of Himachal apples from other regions of the state. Apple harvesting starts during August, and that is the time when stores start filling up and last till October. The selection of apple quality to be kept in the CA store is done carefully as the shelf life of the fruit has to be prolonged for 7–8 months in CA conditions. The technology has given an upper advantage to FHEL as regulation of temperature, Carbon Dioxide, Oxygen, and Humidity has resulted in quality which is almost at par with the fresh apples. This not only resulted in high quality of produce being offered to consumers but also minimization of wastage losses, shrinkage, and scalding problems.

FIGURE 11.4   FHEL facility at Rai, Sonipat, Haryana, India.

FHEL store is the biggest of its own kind in the country with a capacity of 12000 MT storage with 78 chambers. The parameters of ripening control like temperature, humidity, nitrogen, oxygen, and carbon dioxide are computer-controlled.

Apart from this, the others are:

- Imported Italian Grading and Sorting lines.
- Fully atomized washing, grading, and waxing facility.
- Packing machines both for bulk and retail packs.
- Germany based air conditioning equipment.
- Israel technology-based bins.
- 200% backup.
- Entire facility is fully air-conditioned.
- Electric stackers and forklifts inside the facility.

11.11 BLACK BOX TECHNOLOGY

Black box technology is a new breakthrough in cold chain management of fresh fruit that ensures to keep them fresh up to 1000 days, about 65 months, which is even more than the life cycle of a fruit. As in IT terms, Black box technology is well-known technology where input and output are known, but the internal structure, and working is not very well understood for the purpose at hand or may not be supposed to be known for various security reasons.

The company Viztar Agritech, India, has tied up with a Spain-based company, Nice Fruits, to set up cold stores in the fresh fruits field. This technology is a parent technology of Spain developed by local scientists. The technology will not use nitrogen, which is generally being used in cold stores for fruit preservation. The technology has been tested worldwide and patented in the USA. It has been estimated that the capacity of storage of fruits will increase four folds with the use of this technology (Figure 11.5).

11.12 RFID SUPPLY MODEL: SUCCESS STORY OF NINJACART

An agritech startup Ninjacart is another self-built success story in the Indian Agro supply chain model to keep pace with international business, with a delivery accuracy rate of 99.8% all the round year with even a single day off. Behind the success of this Smart supply chain is the team of Ninjacart that strives continuously for no error in the process and controlling pilferages.

FIGURE 11.5 Black box.

The four-year-old Bangalore-headquartered startup began by focusing on the retail-consumer segment (B2C) has created a B2B model to set up a seamless link between farmers' produce and retail stores. The goal was to ensure a fair price for everyone involved (Figure 11.6).

FIGURE 11.6 The supply model of Ninjacart.

Source: https://yourstory.com/2019/03/startup-ninjacart-tech-enabled-supply-chain-farmers-yrlcfr3a40.

Ninjacart works on Backward Extension with set up of more than 20 collection centers close to around 7000 farmers and more than 2000 transactions a daily.

11.12.1 SUPPLY CHAIN DYNAMICS FOR NINJACART

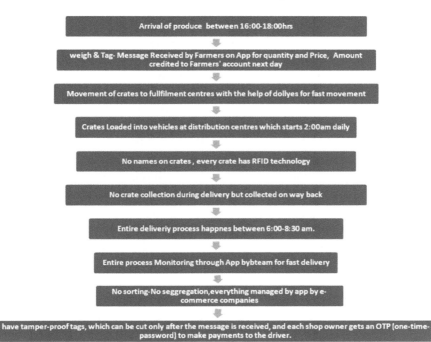

11.12.2 OPERATIONAL CHALLENGES FOR NINJACART

- To ensure timely payments from shop owners;
- Hedging against losses, wastages, pilferages;
- Financial misappropriation;
- Collection of empty crates.

11.12.3 ACHIEVEMENTS FOR NINJACART

- Around more than 5000 deliveries a day;
- Customer base grown more than 100% in the last 3 years;

- Logistics network grown fourfold over the last 4 years;
- Transportation of fruits and vegetables to small town in cold trucks due to non-availability of cold stores;
- Setting up of cold stores apart from supply chain as trading model;
- Achieving 36 hours cycle from farm collection to making empty crates back to farm after delivery and collections.

11.13 DIGITAL SUPPLY CHAIN SOLUTIONS

The technology works on multiple sensors on the internal attributes of fruits. The data collected can handle more than 55 million fruits per day. The technology can take more than 300 photographs per day which means data up to 30 MB is recorded per fruits. A big sorter could produce 1,6-petabytes of data per day. The scope of digitization fluctuates by area nevertheless. In a comparison of digitization by diverse trades, information and communication technology (ICT) and media are the leaders with relatively high digitization, with agriculture at the very bottom of the list. Agriculture falls in a class that is big and localized with low efficiency, which can convert for efficiency and healthier facility distribution.

Digital evidence can open up openings for all key players in the fruit industry, such as growers, producers, packagers, traders, distributors, and retailers. With more organization and categorizing data available, cultivators can apply even more meticulous agriculture in producing their fruit. Attribution and traceability will be much better. Packhouse procedures can be more creative and conservation more predictable. Delivery is more efficient, providing the right transport at the right time, and retail's last fulfillment will be aligned to the product.

precision Agriculture	• sorting data to inform future farming practices.
Packhouse Operations	• optimising yield, throughput, and uptime through machine data.
Distribution	• reducing waste through alignment of packaging and transport.
Retail	• integrating production data with last-mile fulfilment practices

11.14 CONCLUSION

The maximum amounts of losses in fresh fruits in India after harvest are only due to inadequate supply chains. With the advent of digitalization and AI, it has been proven that smart Supply Chain Technologies can change the dynamics to a great extent and bring India's each and every commodity at a competitive level in International Business. These technologies can only help if we turn the scenario first of all at farm level to all farmers by backward extension through latest apps, downloads, information also, we are able to develop a strong chain of our fruit growers who all are linked in this Smart world to the latest International scenario and markets. Technology has the greatest role to play here as in International business there are many mechanisms to slowdown fruit ripening, and we can remarkably bring change in our society.

KEYWORDS

- **CA technology**
- **controlled atmosphere**
- **financial misappropriation**
- **information and communication technology**
- **sensors**
- **supply chain**

REFERENCES

National Horticulture Board of Statistics. http://nhb.gov.in/Statistics.aspx?enc=WkegdyuHokljEtehnJoq0KWLU79sOQCy+W4MfOk01GFOWQSEvtp9tNHHoiv3p49g (accessed on 21 June 2021).

Pandey, M. (2009). *ICT System of Apple Supply Chain.*

The Agricultural and Processed Food Products Export Development Authority. https://apeda.gov.in/apedawebsite/ (accessed on 21 June 2021).

www.imd.org (accessed on 21 October 2020).

www.logisticsbureau.com (accessed on 21 October 2020).

www.popmafruitexpertise.nl (accessed on 21 October 2020).

CHAPTER 12

Conceptual Integration of International Marketing in India

VIKAS GARG,[1] SHALINI SRIVASTAV,[2] POOJA TIWARI,[3] and SONAM RANI[4]

[1]Associate Professor, Amity University, Uttar Pradesh, India

[2]Amity University, Uttar Pradesh, India

[3]ABES Engineering College, Ghaziabad, Uttar Pradesh, India

[4]GL Bajaj Institute of Management and Technology, Greater Noida, Uttar Pradesh, India

ABSTRACT

As the globalization increases nowadays, firms are getting universal and reach worldwide to make their existence. Internationalization motivates the normal firms to sell their product at the global market and increase their spam of business becoming worldwide. Global advertising has made lives simple as the accessibility of merchandise and enterprises are presently being safeguarded to all, in any case where they are living at present. In the event that we see and watch intently, we will find that really global promoting has made our obtaining design more extravagant and expanded when contrasted with past occasions. It is because of global selling/creation of items that our everyday exercises and utilization examples have been rearranged /reclassified. India has been at the epic focus of such a lot of happing and structural moving regarding business sector powers are concerned. Not just purchasers/clients are by and large vigorously profited also from worldwide advertising marvel. There are a few elements of universal advertising with regards to various locales and countries all over the world. The item/administration structures and procedures change, the methodology towards the clients change, all the more critically the whole showcasing ideas are upgraded, remembering the various arrangements of personal conduct standards of purchasers. The vast majority of the world has turned into a solitary biggest market with all most comparative characteristics and universal advertising considers to be countries as one and attempt to continually provide food the unmistakable needs of it. Worldwide promoting is clearly extraordinary with household advertising yet both have huge similitudes too. These likenesses and difference among household and universal showcasing is intriguing mix to see and learn too. The firm which has aptitude in household advertising can utilize it effectively universally, when it develops to that level. Conversely, the firm which has a generous measure of involvement in worldwide showcasing can utilize their worldwide learning in providing food neighborhood clients/clients all the more successfully and with global ability which can fulfill the clients better and can put the organization into a ruling position. The worldwide advertising has an alternate biological system and set of difficulties out and out, there are numerous variables which straightforwardly and in a roundabout way contribute in the achievement and disappointment of universal showcasing. Globalization has been motivating the local or domestic firms to sell their product in the global market and to come out from their space. This pushing is once in a while not invited by firms, and they ward all the while. Actually, Global trading helps normal firms to increase their income and benefits by selling their product to different customers in different markets. The present paper is an endeavor to comprehend the center ideas of global showcasing and every one of the difficulties/openings which are always looked by universal promoting. This paper additionally examines the different angles like likenesses and differences among household and universal advertising. The extent of universal advertising is exceptionally tremendous in India; this paper likewise underlines this specific perspective with appropriate clarification.

12.1 INTRODUCTION

Marketing at the international levels one of the most significant parts of globalization as far as fulfillment of clients/clients. The whole world has turned into a solitary market, firms can't rely on a solitary market any longer, and they bring to the table their item/administrations from one country to another. This selling of product to more than one market calls for showcasing in those business sectors too. Presently the techniques of showcasing for domestic and global market is same but the difficulties increase. The significant point to come each and every mind that why firms are getting more curious to go international market? The answer of this question varies organization to organization. After the globalization began coming to fruition on the planet and especially in India, firms think that it's exceptionally simple to move into any universal region with their institutionalized item/administrations. The common effect of government easy and liberal strategies and open market available also act an important factor for the development of international market. Through this not organization making profit but also customer getting advantages to get different product in terms of quality also. Advanced age has additionally assumed a fundamental job in carrying new thoughts and creative ways to deal with training in getting more extensive client base globally. Showcasing of items and administrations require distinctive level of ability in global space, the customary methodology needs to altered and new ways must be received when managing worldwide degree of rivalry and quality detail. Promoting has moved toward becoming spine of the business exercises and this progress has occurred all around quickly, the fascinating certainty remain that the showcasing is predictable with deals and benefits of firms. Global advertising depends on specific devices in seeking after the consumer loyalty that is showcasing examination and promoting knowledge. Promoting has developed as a significant and powerful apparatus for expanding deals and incomes for the firm and placing it into a superior position than its nearest rivals. Advertising additionally serves the organizations so as to build up an item/administration according to the necessity and requirements of clients. In advertising the client is given the best accentuation and his preferences and inclination are joined into the item and administration's highlights. Promoting is very inverse to selling, where the clients are disregarded altogether and interests of vendors are main priority. Advertising has included the different game principles and support the organization to face its competitors. Presently the firms getting more expert suggestions and follow their expertise strategies to implement the advertising methods. This can be beneficial for both clients and organization.

12.2 METHODOLOGY USED IN RESEARCH

The research nature is exploratory and various sources used for data collection to get a careful thought with respect to different contributions of global promoting. The examination paper is spellbinding in nature too, where idea of global showcasing is dissected and comprehended from different points/measurement conceivable. Various sources of data used for data collection in present research paper, during the accumulation most extreme consideration has been practiced to maintain a strategic distance from any inconsistencies and deluding certainty. The data which has gathered is carefully from auxiliary sources and their validity has been protected as of now because the wellsprings of data are very solid. These are the main sources from where researcher collected the data are as follows:

1. Go through the different books of International advertising.
2. Various research papers on worldwide showcasing.
3. Various online assets and online examination locales.
4. Various journals and research articles and other important information from web on worldwide showcasing.
5. Other online material available on global showcasing.

Every examination has a few targets and this investigation isn't an exemption in such manner. The destinations of the investigation fill in as the beacon for the examination which means goals manage the whole examination to an obvious end result. The standard examination has number of goals and it meets every one of its destinations over the span of research.

The writing audit process likewise helps in discovering the precise thought of the point in hands. Universal promoting has been a subject of more extensive examinations; it has been considered from every single imaginable measurement both in India and comprehensively. The lumps of those unmistakable investigations are given in the accompanying way.

Taneja Girish, Dr. Girdhar, Raj Anand Neeraj Gupta, in their research paper title "Marketing Strategies of Global Brands in Indian Markets" (2012) wrote that "With increasing globalization and international trade, a number of international brands are entering into India which is one of the fastest growing and highly competitive markets in the world. Though, most of the global firms failed to understand the needs of Indian consumers as well as the market characteristics but there are a few of them who have been successful in positioning their brands into the Indian market because they attempt to understand well the needs of target group before introducing a brand into the market. Even some of the most successful brands into day's time had committed several blunders or mistake while initially entering into Indian market. For instance, Kellogg's, McDonald's, LG, Reebok, and Coca-Cola are among such global brands who initially introduced standard products by following standardized global strategies but later realized their mistakes and thus modified their product or services according to the needs of Indian consumers and became successful. This research is an attempt to investigate why some international brands, that are successful globally, fail to attract significant markets in India."

Steenkamp, Jan-Benedict, and Frenkel Ter Hofstede (2002), in their research paper title "International Market Segmentation: Issues and Perspectives," wrote that "International market segmentation has become an important issue in developing, positioning, and selling products across national borders. It helps companies to target potential customers at the international-segment level and to obtain an appropriate positioning across borders. A key challenge for companies is to effectively deal with the structure of heterogeneity in consumer needs and wants across borders and to target segments of consumers in different countries. These segments reflect geographic groupings or group and consist of potential consumers who are likely to exhibit similar responses to marketing efforts."

1. Ghauri, Pervez, and Cateora, Philip in their book title "International Marketing" (2009) wrote that "these forces affecting the international business have led to a dramatic growth in international trade and have contributed to a perception that world has become a smaller and interdependent place."
2. I look at the Swiss Multinational Company, Nestlé, 'The Food Company of the World'; it claims its products are sold in every country in the world. It has factories in more than 80 countries and it has many brands that are recognized all over the world.
3. Toyota and its subsidiaries sell their cars in more than 170 countries, giving it a presence in more countries than any other auto manufacturer."
4. Kozak, Yuriya, and Smyczek, Slawomir in their edited book title "International Marketing" (2015) found that "Increasing integration with the world community, of the domestic enterprises into the foreign markets and intensifications of development of new forms of the international business are the main present tendencies of economy reforming for countries. These countries system of the world economy, and the way this process will take place, efficiency of further economic and social development of the states, as organic sub-economy depends. Efficiency of occurrence in world economics is defined activity of its business structures. Successful activity of the enterprises in the foreign markets is possible only at skillful use of receptions and methods of the international marketing activity."
5. Saxena, Sandeep in his research paper titled "Challenges and Strategies of Global Branding In Indian Market" (2012) writes that "number of well-known global brands have derived much of their sales and profits from non-domestic markets for years, for example, Coca-Cola,

Shell, Bayer, Rolex, Marlbordo, Pampers, and Mercedes-Benz to name a few. Brands such as Apple computers, L'Oreal cosmetics, and Nescafe instant coffee have become fixtures on the global landscape. The successes of these brands have provided encouragement to many firms to market their brands internationally. A number of other forces have also contributed to the growing interest in global marketing. These are the perception of slow growth and increased competition in domestic markets, belief in enhanced overseas growth and profit opportunities, desire to reduce costs from economies of scale, need to diversify risk, and Recognition of global mobility of customers. Today companies going global, continuously innovate their strategies for worldwide success. Global Marketing needs clear vision regarding the 4P's, i.e., Product modification, Pricing issues, Promotion mix strategies to adhere to the cultural sentiments, language, and lifestyle patterns of foreign consumers and right distribution channel to penetrate deeper. Other challenges include suitable Packaging and building Brand for acceptance in the foreign market."

6. Viswanathan, Nanda, and Dickson, Peter in their research paper title "The fundamentals of standardizing global marketing strategy-2007) find out that "The increase in world trade, increasing integration of the world's major economies, and the onward march of globalization, will mean that decisions on standardization and adaptation of marketing strategies will continue to be an important issue for academic research and marketing practice. In spite of the substantial research on standardization/adaptation of marketing strategy for over 40 years, the theoretical foundation for standardization/adaptation research remains weak (Ryans et al., 2003). There has been with the exception of Jain (1989), Cavusgil et al. (1993) and Johnson and Arunthanes (1995) little attempt to develop a theoretical framework that would be informative on standardization issues." Furthermore, all available theoretical foundations of standardization "center on the perception of consumer homogeneity and/or the movement toward homogeneity" (Ryans et al., 2003). "While consumer homogeneity is an important issue, the dimensions of marketing strategy go beyond a consideration of the customer. In particular, competition plays a critical role in the development of marketing strategy and consequently in decisions on the degree of standardization no marketing strategy."

7. Ahmed, Manzoor, Ullah, Shafi, and Aftab Alam in their research paper article title "Importance of Culture in Success of International Marketing" (2014) found out that "It's very important to discuss

about the importance of culture in international marketing, after firstly understanding what an international marketing is and what culture is. Culture is valuable for doing trade in local market but it is more significant for international marketing, there as on being that in international marketing people have different believes, nature, culture, or language. All these aspects create the problem of managing people in international marketing so it is important for any organization to understand the cultural differences before going for business in international markets."

8. Meaning and definition of International Marketing-We can characterize global advertising basically as the use of showcasing standards and ideas to more than one country. In any case, in spite of basic conviction, there is a minor difference what is known as universal advertising and worldwide showcasing, which is accepted to be comparable term, yet it isn't so. There is gigantic distinction which recognizes universal showcasing from worldwide advertising. Global showcasing sees the world in various markets and sections and provides food as indicated by it. On different hands, worldwide promoting considers the to be world all in all unit and does not recognize advertise on any grounds and gets ready items and administrations for worldwide clients as opposed to for anyone country's clients. Universal promoting has essential standards like local showcasing.

According to the American Marketing Association (AMA), "International marketing is the multinational process of planning and executing the conception, pricing, promotion, and distribution of ideas, goods, and services to create exchanges that satisfy individual and organizational objectives."

Definition by experts-The portion on global advertising by the specialist will help to more clear understanding of global advertising. These definitions will help to more clear understanding of different components of universal advertising.

"There is a global approach to international marketing. Rather than focusing on country markets, that is, the differences due to the physical location of customers groups, managers concentrate on product markets, that is, groups of customers seeking shared benefits or to be served with the same technology, emphasizing their similarities regardless of geographic areas in which they are located"—*Muhlbacher, Helmuth, and Dahringer (2006).*

"International Marketing is the performance of business activities that direct the flow of a company's goods and services to consumers or users in more than one nation for a profit"—*Cateora and Ghauri (1999).*

"International marketing is the application of marketing orientation and marketing capabilities to international business"—*Muhlbacher, Helmuth, and Dahringer (2006)*.

"The international market goes beyond the export marketer and becomes more involved in the marketing environment in the countries in which it is doing business" —*Keegan (2002)*.

International marketing is defined, "An application of marketing concepts and principle across two or more nations, for the motive of maximization in sales and market share, International marketing is all in international territories; it an extension of domestic marketing, as most of the concepts remain same."

The above arrangements of definitions, in a route, get out the different attributes and further give an explanation with respect to the idea of international advertising. The definition given above give the accompanying attributes of International advertising:

1. It is only an augmentation of residential showcasing as far as center ideas.
2. It causes firms to exhaust into fresher markets and tap potential outcomes accessible outside of local area.
3. This is an idea which is amazingly submitting face to globalization.
4. High aptitude of skills is required in International Advertising.
5. Maximum companies long for entering into outside market; promotion at international level is one of the most significant apparatuses for doing as such.

12.3 INTERNATIONAL MARKETING AND INDIA: A CONCEPT

As the globalization has ascended and Indian markets ended up being progressively increasingly open to by empowering privatization, the significance of International advertising is on the ascent. Prior to that, In India because of few reasons the international advertising was not very famous. Firms just centered on household clients. After the post-globalization, according to specialist there is a new demand for household industries that is internationalization. Indian Items has picked up a significant spot in world market, new and creative methods for showcasing helped them to get success and open the new door at international level. In India, this eager internationalization has intensely helped in the development and business foundation of worldwide. Earlier that international exhibiting had played an insignificant role for Indian firms, although situation has been changed and continues been changing including new measurements and encounters. Moreover, one could state or watch the accompanying regarding International showcasing condition in India.

The mentalities of firms which are settled in local markets are have been roused to go past the household space because of a few reasons including levels of popularity of Indian items with focused costs:

1. The assistance and impetuses given by regulatory bodies additionally promoted to organization to move ahead than their typical region of working.
2. The development in advertising foundation like media, data innovations and so on has assumed a significant job in communicating the idea of International promoting.
3. Firms working globally, success of such firms depend on advertising. Advertising helps to spread their existence outside the nation boundaries.
4. Progress in showcasing polished methodology and nature of talented individuals in the area of International promoting additionally helped and changed the whole discernment for the internationalization.
5. New wave/ages of business visionaries have developed who think ambitiously and don't have any desire to control to household benefits as it were. They need to wander out and search for more current and better chances. This frame of mind additionally formed the more up to date roads for more prominent use of International promoting.
6. The believe of clients in Indian products helps the Indian firms to enter in International market.

12.4 KEY DIFFERENCE BETWEEN INTERNATIONAL MARKETING AND DOMESTIC MARKETING

International business had an amazingly insignificant impact on India, but now a continue add on changes and new measurements. Let's see the accompanying regarding International showcasing conditions in India.

1. Domestic showcasing alludes to advertising exercises inside the legitimate and topographical outskirts of the country, on different hands International promoting alludes to showcasing exercises

outside of the lawful and geological guests of the countries yet the exercises continue as before.
2. The zone which the residential promoting serves is little when contrasted with International advertising, on different hands International showcasing serves generally enormous region and clients.
3. Government mediation is very moderate in residential promoting yet in International advertising legislatures of including countries become exceptionally interceding and mindful because it included domestic and international issues.
4. Expansion of business activities at domestic level is exceptionally lower and it is one of the constrained for a country, however the expansion of business activities at global level demand high advertising and it include numerous activities.
5. The utilization of innovation in residential showcasing is low and mirrors the household mechanical condition, dissimilar to it International advertising utilizes high and front-line innovation with most recent advancement and Updating to provide food every one of the business sectors it targets.
6. Domestic advertising has moderately okay rate when contrasted with International showcasing as the commonality and past experience of managing business sector powers is useful in relieving the hazard. The risk level in international advertising is higher because numerous activities involved at one time and it became quite difficult to handle them carefully and safely.
7. Requirement of fund in household showcasing stay low when contrasted with International advertising as against it the prerequisite of money ascends with International promoting as it manages numerous nations when contrasted with residential promoting.
8. Research is significant in both the sorts of advertising, yet in household promoting the utilization and extent of showcasing examination is kept to one country alone, on different hands, the advertising exploration is indispensable and multi-dimensional in International showcasing. As it assumes a critical job in social affair and handling of significant showcasing information to settle on suitable choices in more than one nation.

In a manner, we can put it like it, that International showcasing and local promoting are inaccessible kin, who hypothetically play out similar capacities; however, essentially, they are completely different.

12.5 ATTRIBUTES OF INTERNATIONAL MARKETING

Like some other business, the board and universal business developer, International advertising has a ton of abilities to contribute in the development of world exchange and International exchange specific. Global exchange is full of the floods of International promoting and it helps in development, foundation, and expansion of international exchange. Accordingly, International showcasing has numerous qualities which keep it separated from the comparable winning ideas. A portion of the main attributes are as pursued:

- Caters a tremendous market: Unlike residential advertising, International promoting provides food/servers a generally immense market which may be spread to numerous countries at once. This extraordinary element makes International promoting all the more testing and progressively complex in nature. The market grows the new difficulties and challenges also grown-up related to international advertising.
- Inclusive all the wild components: Global advertising companies have potential to cater to grave dangers and vulnerabilities and easily showcase the progress in an interesting manner. Inspite of the fact that there is stiff competition only few companies are able to survive and compete.
- Requires more professional competency and abilities: International showcasing is skillful employments and further requires high understanding and skills to manage its exceptional difficulties. Global advertising is amazingly perplexing and incorporates more up to date difficulties, to manage such unfriendly and at no other time circumstances such sharp abilities are required which are not all that taken care of in residential promoting.
- Deals with enormous and hardened challenge: One of the most significant qualities which International showcasing has is managing overall challenge and it makes International promoting increasingly unique. It isn't so residential showcasing does not manage rivalry, however, the matter of the truth of the matter is the nature, and power of rivalry is very extraordinary and novel. Such degree of rivalry makes International showcasing increasingly extraordinary and high ability required for working field.
- Developed nations have most noteworthy cooperation in International advertising unlike household showcasing; created economies command the International promoting situation. One might say in the event that we make a rundown of top nations engaged with International promoting

activity, it will incorporate most created countries as it were. It demonstrates their purpose behind improvement over immature countries.

12.6 OTHER CHARACTERISTICS OF INTERNATIONAL MARKETING

- Handles and includes large-scale activity;
- Subject to International limitations;
- Carries an ultra-touchy nature;
- Advanced innovation assumes a critical job;
- Requires certain specific bodies;
- Requires long haul arranging.

During the discovery of the fundamental qualities of International showcasing, one could make out that International advertising has assorted attributes than its nearby adversary local promoting. By intently viewing the extraordinary style of International promoting provide us the reasonable comprehension of genuine shade of International advertising.

12.7 IMPORTANCE CUM OPPORTUNITIES OF INTERNATIONAL MARKETING IN GLOBALIZED ERA

The importance and the role of international advertising can be understood completely during the procedure. Globalization needs universal advertising to sell the product into various part of world. Globalization is the channel or medium that helps a firm to showcase its product worldwide. Global showcasing benefit in development of International exchange by advancing it through different advertising strategies. Albeit International promoting incorporates number of significances in today's globalized ear, a portion of those significance are as:

1. For the growth of firms it is important to present its product to large number of clients and firms also get chance to increase income as well.
2. International advertising significant from the perspective of expanding the brand worth as promoting is tied in with marking and situating company's items and its name. Such introduction helps to a firm setting a stage for worldwide.
3. International advertising provides advantages in getting associated with the global world, as it interfaces every one of the clients of firms with one item/administration. It resembles having amazingly

enhanced client base. It additionally helps in growing new and imaginative items too.
4. Plays an important role in opening entry doors for future development of firms in different market segments and different customers in different nations.
5. For the international clients imperative to get the product which are not available in their region and through international advertising they are able to know the various product and also the merchandiser get empower to deliver the product in different region of world.
6. International showcasing utilizes most recent and inventive creation advances which consequently decrease the general expense of the items. This encourages clients to expend items at low costs. It has been seen that sometimes that local items are beloved than the International items.

12.8 BENEFITS OF INTERNATIONAL MARKETING

1. Helps in receiving the rewards of upper hand.
2. Production/Consumption of new and creative items/administrations.
3. Increase in utilization because of immense supply from outside of nation.
4. Increases all out generation of firm hence giving it points of interest of economy of scale.
5. Increases fare income of firm, for the most part acquiring remote trade of high esteem.
6. Challenging normal catastrophes and defeating from it.
7. Knowledge and social advancement among countries and harvesting worldwide harmony and agreement.
8. Country picture improvement.

12.9 PROBLEMS FACED BY INTERNATIONAL MARKETING IN INDIA AND ALIKE COUNTRIES

In every day working the International showcasing has been facing lot of difficult situations. These specific circumstances now and then alluded as difficulties can be open doors also given the sort of initiative the universal firm has. Interestingly, local advertising additionally faces heaps of difficulties in executing its assigned work, yet the difficulties confronted globally are somewhat extraordinary and intricate when contrasted with residential difficulties. A portion of these difficulties are clear in this manner can be met with no much exertion. On different hands, a few difficulties are dangerous and need uncommon/specialists to help to turn out from. A most significant aspect concerning International promoting is that challenges are nation explicit. Besides, changing globalization's scale and extreme moves in customer practices overall just enlarge the test extent of International showcasing. A portion of the main difficulties looked by worldwide showcasing are recorded underneath:

1. **Facing Extraordinary Assorted Variety in Culture:** International advertising needs to manage outrageous enhanced societies and networks, which have their own arrangement of qualities and convictions. The greatest test to deal with the extraordinary discernment and practices of these particular gatherings turns out to be trying for International advertising. Presently universal advertising needs to embrace diverse arrangement of systems and method for execution; it just converts into more issues and testing circumstances.
2. **Very Extraordinary Promoting Condition:** Global advertising companies needs to showcase its position in an incredible manner in all those countries where they plan to sell its items. Every nation with various arrangements of political and legitimate arrangement alongside innovative viewpoints makes International advertising increasingly confounded and consequently become an incredible test.
3. **Different Buyer Conduct:** International promoting counters one of the difficulties as far as various nations having distinctive shopper conduct. The distinction in taste and interests and different viewpoints make crafted by International showcasing somewhat confused than it as of now is.
4. **A Certain Observation towards Outsiders:** Marketing additionally needs to manage a pre-made a decision about attitude of individuals remaining in various nations related with outsiders. Global advertising is finished by remote firms just so they face such bias more. This test has been winning from an exceptionally lengthy time span.
5. **Problem in Choosing Advancement Apparatuses:** As against to residential advertising, universal showcasing manages decent varieties and complexities regarding society/religion/demographical fragments and so forth, picking an advancement instrument or strategies turns out to be exceptionally testing it such situations.

6. **Different Market and Different Powers:** One of the common challenges in International advertising is to manage a number of markets at one time. A global market is the set of various market powers that allude. It required a lot of effort and planning to manage the power of one market and to execute the plan simultaneously in different markets.
7. **Challenge of Managing in Various Markets Simultaneously:** When a firm receives an order from its worldwide clients. It deals in more than one market; therefore, the examination of managing various markets with restricted assets surfaces. Such a test isn't unmistakable in residential showcasing, yet makes a great deal of awkward circumstances for global firms.

12.10 SOME OF THE MAJOR CHALLENGES FOR INTERNATIONAL MARKETING

1. Different perspectives for purchasing decisions.
2. Restriction imposed by government and laws.
3. Rules, law, and regulation related to world trade.
4. Language and social contrast alongside human standards of conduct.

There is more challenging and difficult situations are consistent looked by International showcasing, one thing is certain that it isn't exceptionally simple to remain into worldwide markets and perform everyday capacities like local promoting. Organizations must capacity with alert in such manner and practices of International advertising ought to be performed under the supervision and direction of specialists/gifted individuals.

12.11 HOW TO OVERCOME FROM THE CHALLENGES OF INTERNATIONAL MARKETING?

As far as we have seen that International advertising plays an important role for the firms working in different nation and different market. The challenges and difficulties faced by firms are higher in nature and truly hurt the core objectives of the firms. These types of challenges and problems can be settled down by the team of expert of the field. Apart from this following points can also considered for the similar reason:

1. Global firm ought to build up an itemized and top to bottom comprehension towards the focused-on market with respect to all the measurement and potential difficulties. Firm should show them out and get ready assets and different devices to beat beginning issues.
2. International firms ought to create awareness towards the different countries and their culture of every nation they are working. This exercise can be beneficial for facing difficulties at a global level.
3. Appoint individuals with the good skill set and able to face challenging situations, and these individuals can be beneficial in recognizable proof and managing challenges of International business.
4. Conduct research from time to time to keep updated about the latest and possible challenges in International Market. Such research can be helpful to overcome and ignoring such type of issues.
5. Maintaining individuals to individual's interaction is the best procedure to maintain a strategic distance from difficulties in International showcasing, and this contact can likewise give you direct data with respect to the moment subtleties and particulars in regards to the way of life.
6. Before wandering into International Advertising forms, must do some basic exercise. Such preparation incorporates learning about the promoting condition with subtleties and chipping away at consistency and unions instead of difference.

Aside from previously mentioned stages, one could generally transfer on firsthand encounters and auxiliary research done on specific issues in specific nations. In spite of the fact that the dangers and vulnerabilities are enormous in International showcasing, yet many firms planning to get enter into International business and seeking after. Global promoting may incorporate numerous difficulties and deterrent; however, the manner in which outs are likewise numerous and such vulnerabilities can be made do with appropriate assets in hands and responsibility towards a superior development of the organization.

12.12 CONCLUSION

One could comprehend the significance of International advertising by basically understanding the center issue that organizations can't support for a longer timeframe without getting out from their residential space. In basic terms, on the off chance that organizations need to develop and extend/

broadened, then going worldwide is the only way to move ahead. Apart from this hypothesis, here we have numerous different speculations that propel the organizations to out of their household arrangement; during this whole research article, a portion of the conspicuous reasons have surfaced because of which firms get International. In the practical world, the journey to enter into the international market starts with the thought of enter into international business. Organization wants to go for global market woks on techniques to enter in global market. This idea and comprehension of International showcasing is very dubious for new entrants yet existing firms consider this to be only an expansion of their residential advertising draws near. With the development in globalization and industrialization, International business has turned out to be extremely essential and nearly simple to attempt. The firm is presently very associate and submitted towards growing themselves into global domains for receiving rewards and tapping benefits/gains. As residential area here and there demonstrates to be deficient for survival of in excess of specific quantities of firms. Once more, it is one of the numerous reasons why firms get worldwide; however, one could underline here that the purposes behind getting internationalized are numerous and each firm has its own time allotment and mentality to choose over how and when to go global. The present research paper had numerous destinations and purposes which were firmly connected to tell the whole picture of International promoting. During the examination it is additionally seen that International marketing has diverse arrangement of difficulties and focal points also. Yet, one most important thing for future development, firms must go out national boundaries.

12.13 FINDINGS OF THE STUDY

1. International promoting is the way toward applying advertising standards over the local space of a nation for indistinguishable reasons and purposes and destinations from of household showcasing. This is quite challenging task to take thoughtfully international advertising does not contrast from residential promoting.
2. In the era of industrialization, enterprise, and globalization; International business has become just and with universal business developed International promoting.
3. Firms which particularly have a place with India and other creating countries have grown a ton in universal business; such nations additionally assumed the job of host in all respects effectively.

4. Due to constant development and advancement in international business strategies, expert of this field have done great job for making the procedure of internationalization as smooth as conceivable.
5. There is no uncertainty that the procedure of internationalization is extremely testing and brimming with vulnerabilities and represents a ton of troubles as we have seen the rundown in this exploration paper. Be that as it may, with appropriate arranging and definitive authority such difficulties can be met effectively as the recommendation-recorded show.
6. Countries like India Internationalization or international showcasing is very helpful because countries like India have items which can be sold worldwide at very affordable prices. The global business helps us to increase production and also increase the workforce. These two explanations are very much in favor of support of international promoting.
7. The inception of present research work remembering certain goals, every one of the targets has been satisfied with legitimate evidential arguments. Innovation and other advertising natural elements are assuming an exceptionally vital job in the development and enhancement of organizations from domestic to global market.
8. The degree of jobs and consolation the legislature of India has been offering to firms who wish to get into the worldwide market has changed and developed in the most recent couple of years, and this has reflected upon the ongoing momentous development of Indian source firms working universally and included into a global business.

KEYWORDS

- **customers satisfaction**
- **ecosystem**
- **global business**
- **globalization**
- **international marketing**
- **internationalization**

REFERENCES

Ahmed, M., (2014). Importance of culture in success of international marketing. *European Academic Research, I*(10). ISSN: 2286-4822.

Jan-Benedict, E. M. S., & Ter, H. F., (2002). International market segmentation: Issues and perspectives. *International Journal of Research in Marketing*.

Kozak, Y., & Smyczek, S., (2015). *International Marketing*. ISBN: 9786610601719.

Saxena, S., (2012). Challenges and strategies of global branding in Indian market. *IOSR Journal of Business and Management (IOSRJBM), 4*(1). ISSN: 2278-487X.

Taneja, G., Dr. Girdhar, R., & Gupta, N., (2012). Marketing strategies of global brands in Indian markets. *Journal of Arts, Science and Commerce, III*(3(3)). E-ISSN: 2229-4686. ISSN2231-4172.

Viswanathan, N. K., & Dickson, P. R., (2007). The fundamentals of standardizing global marketing strategy. *International Marketing Review, 24*(1). Emerald Group Publishing Limited.

CHAPTER 13

Blockchain Technology and Its Utilization in Tracking Milk Process

ANITA VENAIK,[1] RICHA GOEL,[2] and POOJA TIWARI[3]

[1]*Professor, Amity Business School, Amity University, Noida, Uttar Pradesh, India*

[2]*Assistant Professor, Amity International Business School, Amity University, Noida, Uttar Pradesh, India*

[3]*ABES College, Ghaziabad, Uttar Pradesh, India*

ABSTRACT

Blockchain technology is a complete secure distributed database (we also refer to it as a distributed ledger) where data is stored in every node. All stakeholders are equally responsible for every transaction. It gives rise to high security in the digital transaction, so many large companies are already changing their infrastructure adopt BCT, to name a few companies like McLane Company, Nestle, Unilever, Tyson Foods IBM Food Trust consortium including Dole Food, Driscoll's, Golden State Foods, Kroger, McCormick, and Co., has also been examining blockchain technology for the food supply chain and secure traceability. Even China-based company Dianrng and FnConn have already launched successful chained finance to enable supply chain using BCT.

13.1 INTRODUCTION

To enhance integrity and consistency in transactions, blockchain a distributed system that runs on multiple nodes across multiple data centers where transactions are accumulated in an encrypted "block" with new entries added as they occur to make a "chain." Each block contains a cryptographic hash of the previous block, a timestamp, and transaction data and is available at all nodes. Copies of this block of transactions are stored in multiple nodes, and each node's data is updated simultaneously on a real and virtual basis. If any breach is created, it will be intimated as tampering with data is next to impossible because it works on group consensus. A blockchain is actually a database because it is a digital ledger that stores information in data structures called blocks; these blocks are compared to determine if one has been tampered with, and only the consistent blocks continue to be used. Each block provides a "single version of the truth" about transactions and activities occurring across complex supply chain ecosystems.

Every technology work on some architecture as it is the backbone of any enabled technology; hence because of seamless integration and rise of distributed network systems, the blockchain works on POW (proof of work) consensus model, in which all the nodes get the right to publish the next block by solving a computationally intensive puzzle but it requires high computational power of computers and high connectivity. If implied properly, the result of the computation is easy to verify and it will also help other nodes to validate and update the blockchain easily. The beauty of the game is that the node which solves the computational puzzle receives a share automatically and this process is called "mining" and person is called the miner who mines the transaction. Because this is time and energy intensive, other alternative methods of verifying a block, such as proof-of-stake, have been developed and implemented in subsequent spin-offs stakeholders. The Idea Behind implementing blockchain is to know your customer better and serve them well and KYC. Blockchain in KYC can be proved as one of the most promising applications of the decentralized technology, serving a real need by decreasing KYC administrative costs and lost time while at the same time increasing security and transparency.

BCT can not only access and inspect, but also can and add to the data each time which will be updated on real time basis. With the encryption technique which prevents them from altering or deleting existing data. The original information stays permanent, leaving a integral private and public information trail of transactions. Blockchain implementation requires many factors like: (1) Complete data provenance (2) open standards for infrastructure (3) Architecture which is distributed (4) Agreed governance among all stakeholders.

13.2 SUPPLY CHAIN AND BLOCKCHAIN

Supply chain management is not new to the industry, it has helped organizations to have optimal solutions and just in time inventory, but still there is lot of mismatch between the actual and predicted transactions in the real world. Supply chain defining expected good practices and compliance requirements across areas such as food safety, quality, environmental sustainability, provenance, etc. If we use blockchain technology in the supply chain it can further improve quality, cut down the cost, adding the utility to the product and services to enhance customer value. The customer is always looking for innovation in product they buy which has a fresh, palatable, nutritious, and safe reach along with the choice of their connivance and pricing as per their expectations. Organizations are looking for a solution where there is connectivity from farm to table. Companies wanted to keep upright in this competitive world keeping traceability of product at various stages, doing quality tests to detect and prevent food hazards due to contamination and give customer a value addition, and remain competitive and sustain their business. Here focus is milk turned in skimmed form.

13.3 A LOOK COMPLETE TRADITIONAL FOOD CHAIN

The milk food supply chain is an interwoven complex system which can be broadly represented as the different divisions like primary and secondary changes.

As described above Primary producers' of milk which can then undergo a variety of processes before it reaches to its final destination, which is end user through a range of sales channels. It will go through multiple stages processing, sub-processing, and will combine ingredients to change its various stages, doing all the quality of the product has to be maintained.

Secondary processing changes the liquid state of milk into a state where it can then be used as an ingredient, or can be sold as a final product directly to consumers, like skimmed milk. Manufacturing creates more processed products which have undergone several changes. It requires a seamless platform; IoT enabled technology, and private track and trace repositories to achieve a more complete solution. With the private track and trace repository and blockchain management applications, companies can select what information and to whom they want to share it along the supply chain. Sensitive information is kept private blockchain technology with the help of IoT is having potential to support a secure 'end-to-end' view of the data across supply chains.

If we integrate IoT (Internet of things) with blockchain technologies using sensors and RFID Tags. Data which is sourced from a validated and trusted 'origin' will be recorded. We cannot deny the fact that The food supply chain incorporates a plethora of technologies and devices such as mobile devices; smart sensors on storage facilities or transport; smart packaging; surveillance equipment or detectors able to determine food product composition in a non-destructive way all transactions will be recorded and will be available to all process owners in a real-time basis.

Understanding how to verify the provenance and source of data stored on a ledger needs to be investigated as part of any research conducted in this space.

The expected features of an effective value chain implemented will be as under:

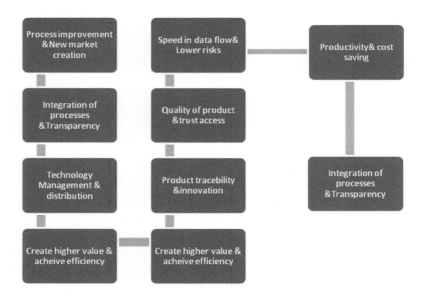

Adoption of BCT will help us to get integration among various components; it will remove middleman, and the transaction will be directly between the seller and buyer. It will incorporate a fully auditable and valid ledger of transactions, which is unforgettable and indelible. For making changes in the ledger, every individual blockchain in the system needs to be altered; hence fraudulent transactions can't be added, and it is impossible to delete a blockchain transaction in an attempt to hide it. Hence every transaction will be independent yet interdependent and highly secure, and also transparency will be in all transactions.

13.4 PROPOSED MODEL OF SUPPLY CHAIN USING BLOCKCHAIN AND IOT ENABLED TECHNOLOGIES

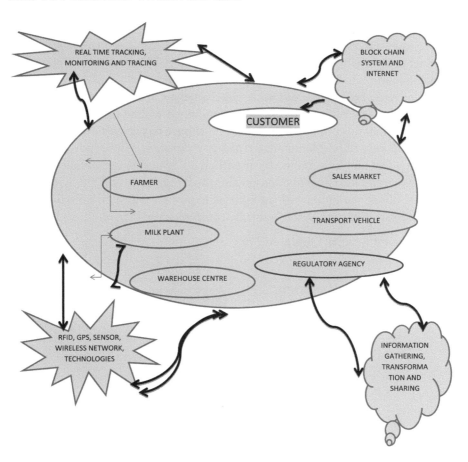

The diagram proposed above will improve supply chain transparency, traceability, and reduce transaction costs. Using blockchain technology having cryptographic trust, which is not only inherently traceable but also time-stamped and sensor resistant, distributed ledger and is near real-time too. BCT will involve algorithm-based smart contracts, which will enable the integration of contractual relationships between trusted trading partners through self-verifying and self-executing agreements replacing intermediaries completely; hence it will cut the costs, maintain security and connectivity.

It signifies how the supply chain traceability system established, which relay on RFID technology to handle data acquisition, circulation, and sharing in production, processing, warehousing, distribution, and sales links of skimmed milk supply chain relating to product movement.

BCT will handle all aspects like quality, compliance, contractual, and accounting through a distributed ledger process, which will add traceability and transparency as well. The smart contracts enable self-establishing contracts among the stakeholders will improve trust among the stakeholders. Smart contract will help to eliminate errors and avoid duplication and unnecessary reconciliations, which will lead to improved efficiency and lower down the cost. It will provide holistic collection, storage, analysis, and insight solutions and emerge as a successful model as all changes made will be recorded in a block where immutability is there, and there is no chance to play with the records, as all transactions will be interconnected and every process will be independent yet interdependent and interconnected. It will give a seamless integration and flow of processes and lead to quality production, cost reduction, and timely delivery; the shelf life of the product along with its expiry all will be on the tips of all stakeholders, which will lead to a satisfied customer and high return on business.

Improved Supply Chain will be:

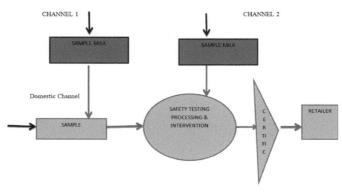

Domestic Channel

13.5 CONCLUSION

With the reinvention of technology, we can reengineer the whole process of the supply chain, which will help to mitigate risk, increase productivity, save time, and cutting down the cost, and it will help to maintain the database as an open distributed ledger which is immutable, having integrity and is available on virtual and real-time.

We can conclude by saying that blockchain and digital edge technology do have functionality that is useful in enabling global food security with transparency, consistency, integrity, and availability of information anytime, anywhere. These technologies can facilitate distributed and secure digital identities and so as part of an information architecture incorporating secure smart devices on the packaging, in logistics operations, in detectors, etc. If applied properly, these applications of permissioned distributed ledger technology could contribute towards enabling global food security. It will help eradicate the middleman and help the process to be smooth and integral.

KEYWORDS

- blockchain
- digital ledge technology
- distributed database
- food security
- internet of things
- secure traceability
- supply chain

REFERENCES

Blockchain Technologies in Agriculture and Food Value Chains in Kerala; Kerala Development and Innovation Strategic Council (K-DISC), https://www.researchgate.net/publication/324861994 (accessed on 21 October 2020).

Food and Agricultural Organization of the United States. file:///C:/Users/Admin/Desktop/ca2906en.pdf Received: 6 May 2020/Revised: 9 June 2020/Accepted: 10 June 2020/Published: 15 June 2020 (https://www.mdpi.com/2076-3417/10/12/4113/htm) (accessed on 21 October 2020).

Gerard, S., (2017). *E-Agriculture in action: Blockchain in Agriculture – Opportunities and Challenges*. https://beacon-h2020.com/hub_info-en/e-agriculture-in-action-blockchain-for-agriculture-opportunities-and-challenges/. (accessed on 21 June 2020).

Iansiti, M., & Lakhani, K. R., (2017). The truth about blockchain. *Harvard Business Review, 95*(1), 118–127.

Kwon, I. W. G., & Suh, T., (2004). Factors affecting the level of trust and commitment in supply chain relationships. *Journal of Supply Chain Management, 40*(1), 4–14.

Mulligan, C., Zhu, S. J., Warren, S., & Rangaswami, J., (2018). *Blockchain Beyond the Hype: A Practical Framework for Business Leaders*. World Economic Forum, Cologny/Geneva.

Nakamoto, S. (2008). "Bitcoin: A peer-to-peer electronic cash system." Available at https://bitcoin.org/bitcoin.pdf, retrieved 06/15/2018. (accessed on 21 October 2020).

Nakamoto, S., (2008). *Bitcoin: A Peer-to-Peer Electronic Cash System*. Available at https://bitcoin.org/bitcoin.pdf, retrieved 06/15/2018 (accessed on 21 October 2020).

Panta, R. R., Gyan, P., & Jamal, A. F., (2015). *A Framework for Traceability and Transparency in the Dairy Supply Chain Networks*. Aligarh Muslim University, India BABV-Indian Institute of Information Technology & Management Gwalior, India. www.sciencedirect.com (accessed on 21 October 2020).

Petersen, M., Hackius, N., & Von, S. B., (2017). *Mapping the Sea of Opportunities: Blockchain in Supply Chain and Logistics*. Kuhne Logistics University, Hamburg.

Pilkington, M., (2016). Blockchain technology: Principles and applications. In: Olleros, F. J., & Zhegu, M., (eds.), *Research Handbook on Digital Transformations* (pp. 225–253, Cheltenham, UK: Edward Elgar Publishing.

Stadtler, H., (2005). Supply chain management and advanced planning—basics, overview and challenges. *European Journal of Operational Research, 163*(3), 575–588.

Tapscott, D., & Tapscott, A., (2017). How blockchain will change organizations. *MIT Sloan Management Review, 58*(2), 10–13.

The trust machine: The technology behind bitcoin could transform how the economy works. The Economist, 10/31/2015. https://www.economist.com/leaders/2015/10/31/the-trust-machine (accessed on 21 June 2020).

CHAPTER 14

Impact of New Technology on Business

MONIKA SHARMA[1] and VIKAS GARG[2]

[1]GL Bajaj Institute of Technology and Management, Greater Noida, Uttar Pradesh–201306, India, E-mail: sharma.monika1785@gmail.com

[2]Associate Professor, Amity University, Greater Noida Campus, Uttar Pradesh–201308, India

ABSTRACT

Nowadays, technology is helping business organizations to produce more at low cost and get desired results. However, to cut the cost and increase productivity, technology is accelerating the business ability, by using automation, robots technology that replacing the human effort. To understand the changes in work, productivity, and employment, organizations must develop insight into the interrelationship of technology with artificial intelligence (AI), big data, and the internet. This chapter will focus on how organization are making capable themselves to reduce the cost and analyzing the changes in the work and ability or skills of an employee. It will also focus on how technology impacts the employee's performance will help to pursue their creative goal into a new work/life balance technology environment. It is also emphasis on how to change the business strategy (BS) so that organization goal can be achieved by integrate the technology.

14.1 INTRODUCTION

Adoption of effective use of IT in the organization results to increase in the productivity of an organization. In the ERA of IT, to enhance the productivity in the organization, existence of technology alone is not only responsible. In the organization, employee needs to shift their skills toward more abstract and cognitive skills. To develop the business idea, artificial intelligence (AI), robot, and internet helps to organization to create the programs and set the skills.

In a global and dynamic world technology has become foundation to the business. AI, IoT, and robotics are helping to the employee to do work better and faster, but that demanding changes into their skills set or the ways to complete the work in the organization. Technology is also helping to create value and define way to work and how to communicate in the organization.

The groundbreaking, new technology cannot appear overnight. Adoption of technology required some other development in technology, and it took decades of effort to see the effects of technology on work and organization. Organization needs to understand the application of human, physical, and other resources to develop the business.

Advances in Technologies are transferring work into automatic and making easy for an employee or worker to automate their knowledge to complete those tasks which they were thinking impossible and impractical task. Advancement in technology can make it possible to doing things on a large scale instead of doing a small scale. To take out the creative and new insight in a way that evolves the change into market, organization, productivity, and more. The collision of innovation and technology towards industrial revolution, business organization knows how they can transform their business through adopting the technology.

14.2 TECHNOLOGY AND BUSINESS

Technology has good impact on business organization as it has make everything more convenient. It plays an important role to understand the customer needs, improve the business process by anticipating situational tools, optimal use of documents for work, and respond to an instance change in the environment. It brings sufficient value that is not related with labor substitution. Thus elicitation the output from the technology demand alteration in the entire process instead of changes in single process. It is required that business organizations must focus on competitor strategy to ensure no troublemaking into an existing business model (BM).

Technology has emerged an enormous growth in a business organization. It helps business organizations to reform the existing business process and work. The technology has figured out the new business process and way to do the business in a more effective manner. It has given effortless way to perform work more efficiently, faster, and more conveniently. According to Ford, Automation will reduce the cost, can perform work speedier than human, but human needs to learn about the skills to learn about the automation.

14.2.1 ADVANTAGES OF TECHNOLOGY IN BUSINESS

- It is helping to the business organization to be more competitive into market.
- It helps to increasing the productivity.
- It collaborate the people.
- It helps to store the data and information can be access anytime by anyone in the organization.
- It enables new learning and skill set for the growth and sustainability.
- It save time, money, and cost.
- It helps to business to identify the problem and take the corrective action on time.
- It required privacy concern as data or information can be hack by hacker. Therefore, organization needs to be relying on expert.
- It complete the task speedy and on time.
- It helps to create the potential of the employee.

14.2.2 DISADVANTAGES OF TECHNOLOGY IN BUSINESS

- It replace the personal or face to face interaction.
- The lack of information about how to use technology can ruin the work.
- The training, software, and maintain to them become quite expensive for the organization.
- It enables the unemployment and differentiates wage payment.
- It develops sense of insecurity about losing the job.

14.2.3 EFFECT ON WORK AND ORGANIZATION

Technology has been changed the working style and condition of the organization. It has persistently changed the work environment and develops a new way to perform the task or work. It has improved the work process by streamlined the wasteful process and increased productivity (Figure 14.1).

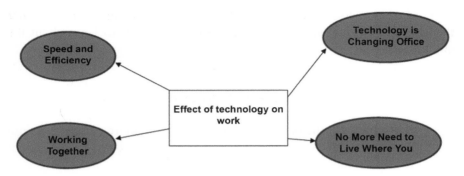

FIGURE 14.1 Technology change benefits to work.

1. **Speed and Efficiency:** Technology has helped to increase the productivity of the employee in all type of organization, i.e., manufacturing, and in communication. The rate of work, production, and speed epidemic increased at which business takes place. Today due to technology the work earlier taken an hours can complete in a single minutes. Timely communication and information is being possible to perform the work.
2. **Working Together Made Easier:** Technology made coordination easier among the employees. Through the online communication tools like video conferencing employees can have meeting with other employee at different place, and the Google Drive tool can use by an employees to share their work to the other platform. Therefore, companies are using workplace management tools to investigate the performance of the employees on some identified project. In the marketing organization, Technology like AI messaging tools are using to follow up the whole conversation that happened between the employee and customer.
3. **Technology is Changing Office Culture:** To create a strong organization culture to tempt in importunity technology has enabled the stereotype changed in workplace by introducing open office space with video game and beer on tap. Beside open office space, organization is also offering freelancer and telecommuter jobs by giving them an opportunity to work from home.
4. **No More Need to Live Where You Work:** The employees who are looking for an opportunity in any company and want to work remotely, they can simply apply on the internet as freelancer. It is a technology who make possible to communicate and work together in a team while you are working remotely.

It helps to both job seeker to work remotely and hiring manager to get talented from anywhere. Technology has also contributed into the social reform and in the advancement of social, cultural, and organizational development. Technology has three core infrastructures: the Agricultural Era, the Industrial Era, and the Digital Era. In the agricultural era, people were dependent on natural resource. In the industrial Era people were focusing on to use the industrial resource by procuring the raw material from other business organization, producing maximum output with minimum utilization of the physical utilization at low cost. The focus of industry was on to increase the efficiency by producing extra unit. In the digital era, people are relying on digital information and knowledge to produce the product and service and defining the business process. This era is based on an infrastructure comprising information and communication technologies. This era enable the business organization to work faster by collaborating and coordinating the task. This era is also called the development of computer and IT and communication Era.

To support the business and organizational activities, organization linking technology and work process (such as enterprise resource planning, customer relationship management (CRM), supply chain management, material requirement planning, human resource management, and enterprise-form automation systems) together.

14.2.4 TECHNOLOGY AND BUSINESS STRATEGY (BS)

How technology can be linked with the BS to implement the plan can be understand by Figure 14.2, approaches to business and technology (Flyod, 1997).

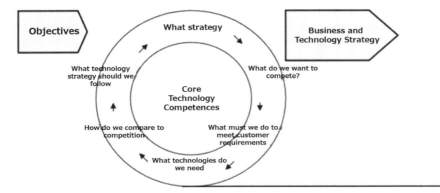

FIGURE 14.2 Structured approaches to business and technology strategy.
Source: Floyd (1997).

The organization needs to develop the objective to link the BS and technology. Organization objectives must be measurable and must to equate within the time. To set the objective organization needs to assess their levels, i.e., corporate, functional, and business levels as they are responsible for the overall organization to develop the strategy and to gain the competitive advantage. Today organization is focusing on technology because it is helping to link the BS with environmental changes and adaptability. Burgelman (2001), stated that technology is also defined to the one business function of the organization.

Vernet and Mohammed (1999) derived a model to understand the relationship between technology and BS and the gap between them (Figure 14.3).

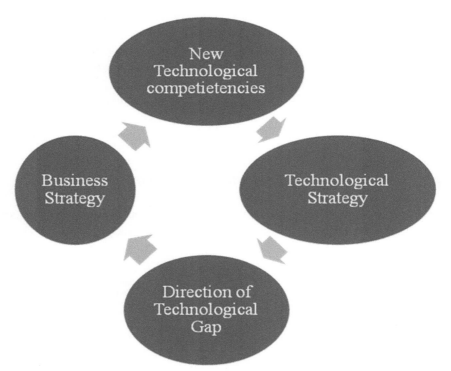

FIGURE 14.3 Two directions in bridging the technology strategy and the business strategy. *Source*: Vernet and Mohammed (1999).

According to this model, today organizations are looking for the environment to initiate competitive effort through adopting different strategically approaches. Technology is helping to create new BM to derive the new business opportunity.

Adoption of the technology is demanding more analytical skills and training to learn about how to use it and to know how to integrate the technology to the other business operation. To get the extract the better understanding about the technology, organization need to explore various strategy possibilities, evaluate them and then revised (Figure 14.4).

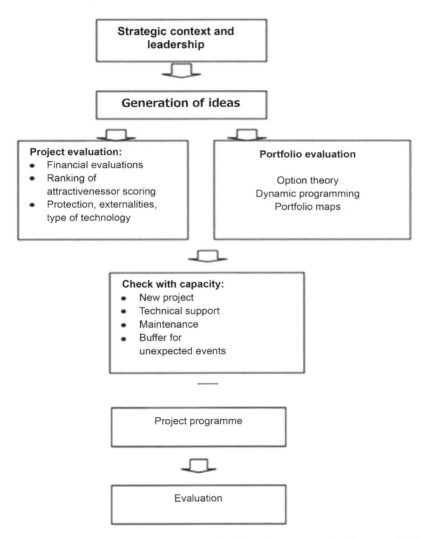

FIGURE 14.4 Explain the framework for organization to determent and implement technology strategy.
Source: Meyer (2008).

Meyer (2008) stated that clear direction and goal are required for technological environment development. Therefore, that organization can be flexible and ready to adopt new technology by evolving the required changes into the organizational strategy. It explains that to implement the technology, an organization must create an environment that can unfold the learning opportunity. So, that new ideas can be possibly generated. Merely organization could have different ideas to implement, here technology help to evaluate each and every ideas to check their feasibility (Figure 14.5).

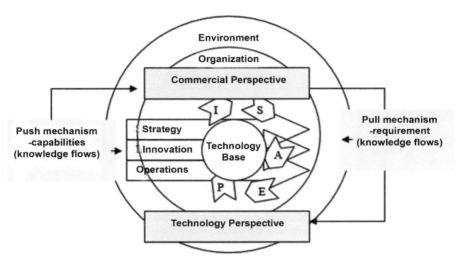

FIGURE 14.5 Technology management framework.
Source: Farrukh, Phaal, and Probert (2003).

Farrukh, Phaal, and Probert (2003) stated that technology management requires comprehension and reservoir based concept to understand the process. The objective of this model is to know how technology supports to business environment, i.e., internal, and external environment. It has defined five steps of technology process such as identification (I), selection (S), acquisition (A), exploitation (E), and protection (P).

14.3 TECHNOLOGY AND PRODUCTIVITY

In today business environment, technology has become great worth for the organization. Technology can help to increase the productivity, the efficiency, improve communication and collaborating the workforce. However, it is an important to choose the right technology, so let us understand how an organization can choose the right technology for their business:

1. Identify the need of the company;
2. Enabling technology required proper training and education;
3. Communication improvement through communication;
4. Improving efficiency through technology.

Productivity helps to improve the living standard of the employee. It doesn't matter how long you are working and how much resources you are using in the organization. Therefore, technology make it possible to increase productivity and evolve new techniques for production. Productivity is measured by the GDP, define the product or production. However, without increasing the output, the technology progress can be increased. For example, if Wikipedia and a free GPS mapping app replaces by a paper encyclopedia and stands along with GPS, the people will be more advantageous even if there is any change in output.

According to the U.S. government agencies, the U.S. is IT producing as well IT producing market, but during last 10 years productivity is going down, statistics said that low emphasis on IT Producing and It using market it tends to decreasing the productivity. To increase the productivity organization has to focus on development of technology and advancement. Figure 14.2 is defining the ways of technology to increase the productivity. There are the following ways of technology to increase the productivity (Figure 14.6):

1. **Connect Remote Workforce:** Due to the demand of contractual or home-based job opportunity, business organization is evolving virtual offices to work collaboratively, resulting to reducing the cost of setting up and maintain the large workspaces. It is being possible for talent-employee to access any information, data, and can take the work status from their employee without meeting them. For example, making WhatsApp group of the employee helping to employers to communication, delegating task and ask the status of work.
2. **Enable the Communal Tools:** There are several tools that are connecting the employee so that collectively they can work for the organization work. For instance, Google's cloud-based application is using in most of organization to share the data and integrate the employees.
3. **Get Organized:** ERP is systems where information about leaves, salary, work, purchasing, and so on stored in software by any

employee of any department, and this could be accessed by any employer without physically meet with the employee.
4. **Be Reachable:** Technology is making possible for the organization to connected their employee by provide updated mobile devices and support system as if any call missed by the employee can cause to the business loss. Therefore, to maintain the steadiness of professionalism the company is using technology like voice mail, softphones, and work caller ID.
5. **Know Productivity Challenge:** Companies are more concentrating to make an evaluation of their productivity challenges so that they can evolve changes in technology tools causing the slow down the productivity of an employee.
6. **Explore Virtualization:** Virtualization helps to the organization to increase business sacking. It extends the lifespan of older computer desktop. It also helps to reduce the cost of the business.
7. **Trust the Experts:** Organization must be distributed the work among the employees and must not be rigid to have an expert for those tasks which can't be automated.

It is advisable that to analyze the work quality and productivity, organization may not be relied on cloud computing as it is dependent upon the technological environment measurement. It will include the change in the existing business process and workforce. The organization needs to be more efficient to learn how to use technology instead of depending upon the invention and formation of the latest and improved technology.

14.4 TECHNOLOGY AND HUMAN WORK FUTURE

Today, organization needs to use the time clock technology to evaluate the way contribution of the employees. In the past time employees were paying based on what they were producing but in the technology era work has been divided into specific task. It is stated that from learning perspective technology is making workplace more flexible and growth perspective. It is discussed about the global labor market, gig economy and how technology is defining the more possibility in the career and job opportunity making free employees from the repetitive and monotonous task. Technology is nor good or neither bad; its matters how we use it. There are several reasons why organizations are investing on work technology:

Impact of New Technology on Business

- Inhuman task can be possible to perform by technology. For example, if in manufacturing companies where work has to perform in extreme hot place, it is not possible for the employee to perform task. Therefore, technology helps to employees to perform that difficult or impossible task.
- Technology helps to reduce the cost of employing the staff which covers almost 50% of operating cost. It demands lowering the price and invest more on new technology that would create a new task for the employee or increase the job opportunity.
- It enables to increase the potential of the employee, helps them to communicate with others, inbuilt the new skills and new process to perform the task.

FIGURE 14.6 Seven ways of technology to increase productivity.

In fact, technology impact on work as it depends on how much organization is spending to adopt the new technology to increase the potential of an employee. There are two visions consider by an organization:

a. **The Fanciful Future:** Imagine if there is no way of recruitment, layoff, retirement, no organization hierarchy and employee not having any sense of stability and employment but employees are collaborating and are productive in technological era as they are acquired the required skills.
b. **The Stipendiary Future:** Imagine that employees are categorized in skilled and unskilled employee and skilled employee are enjoying the job as they are in demand because they have knowledge about the new technology. This world will be referred never ending competition era.

Out technology has created the both fanciful future and stipendiary future. In the fanciful era, organization are using the technology, giving training to develop the potential of an employee about how to use the technology. However, we are also seeing the stipendiary future also which is showing the wage gap and job differentiation and unemployment. The companies are investing more on new technology to make their employee potential.

Technology is also creating the future where employee are not worried about to get job for their survival country it is also replacing the employee work. To change the work of world technology is more fanciful. The impact of technology on work can be understood by discussing the following questions:

1. How an organization supporting to the current work to refurbish the epidemic increase in the output with cost effective? How an organization will support the future work to get more output with cost effective?
2. What role, training, and skills would be required for the today and future job change?
3. Does the current workplace support to adopt the technology? What change would be required to sustain the technology in the organization?

14.4.1 REVALUATE THE POSITION OF TECHNOLOGY

According to, Savvy business and technology executives, the position of technology needs to redefine to gain the competitive advantage and to achieve the business objective. "Because technology advancement demands changes into the role of technology in the organization functions," stated by Satish Alapati, CIO of Media and Entertainment Customer Experience at AT&T. The scope of technological work needs to change due to Business innovation and epidemic. The epidemics can arise from competitor technological BS.

Impact of New Technology on Business

According to the Deloitte's (2018) global survey, to merge the technology and BS together, organization still scuffling with the residual thinking about IT as embrace rather than result oriented.

Many organization believe technology change are important for the BS but somewhere it is failing to understand the impact of technology on work, workforce, and workplace. To understand the technological changes, organization should extend the traditional IT to sustain the work of an organization. The traditional change to the modern IT work is shown in Figure 14.7. The technology has spread in overall the organization by extended their role in the organization.

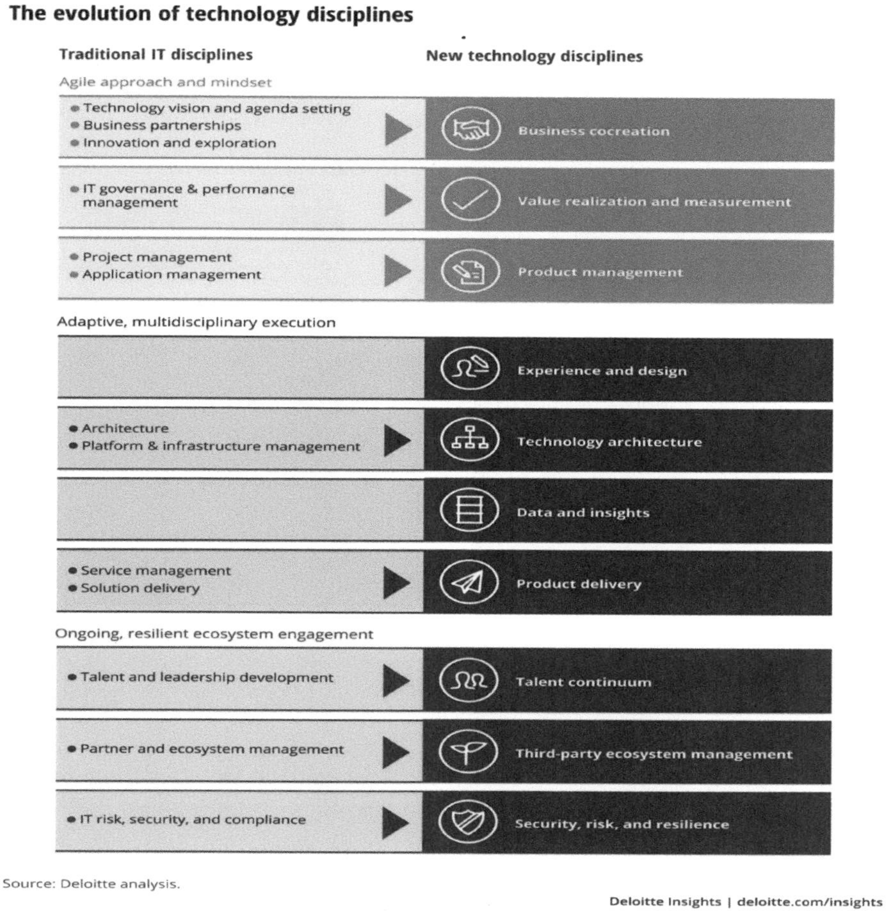

The evolution of technology disciplines

Traditional IT disciplines	New technology disciplines
Agile approach and mindset	
• Technology vision and agenda setting • Business partnerships • Innovation and exploration	Business cocreation
• IT governance & performance management	Value realization and measurement
• Project management • Application management	Product management
Adaptive, multidisciplinary execution	
	Experience and design
• Architecture • Platform & infrastructure management	Technology architecture
	Data and insights
• Service management • Solution delivery	Product delivery
Ongoing, resilient ecosystem engagement	
• Talent and leadership development	Talent continuum
• Partner and ecosystem management	Third-party ecosystem management
• IT risk, security, and compliance	Security, risk, and resilience

Source: Deloitte analysis.

Deloitte Insights | deloitte.com/insights

FIGURE 14.7 Traditional IT to new technology.
Source: Deloitte's analysis.

14.4.2 THREE INTERRELATED DIMENSIONS: WORK, WORKFORCE, WORKPLACE

The technology on work has redefined into three parts: work, workforce, and workplace as stated in Figure 14.8.

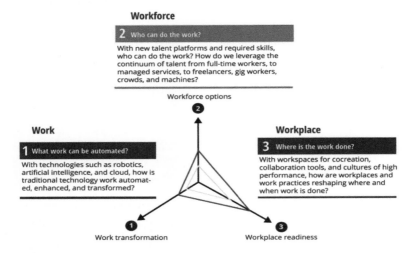

FIGURE 14.8 Redefining the future of work.
Source: Deloitte's analysis.

The future of work of an organization will be depending upon the BS, environmental factors, and competitive advantages forces. However, organization needs to redefine the work and the possible result before the planning about the workforce and workplace.

- **What is to be Done? (Work):** It might be happened that the organization productivity will be increase due the technology and that would be performed by the various factor of production of the organization. However, organization should focus on how to position or arrange the work to get the output. Technology increases the potential of the organization by introducing the process outgrowth focused practice. It has given more meaning to the work of an organization by making

easy to repetitive task which require more effort and skills, resulting to adopting the business technological oriented strategy.
- **Who will Perform the Work? (Workforce):** Technology is demanding to the organization structure as it would require new skills and training. Organization can attain the outcome if they have specialized and trained technological employee.
- **Where to Work? (Workplace):** Technology has been changed the workplace from traditional offices approach to collaboration. Now, because of technology, organizations are giving an opportunity to work remotely by collaborating and integrating or through virtual office. There are no barriers to geographical or vocational.

14.4.3 FORCES SHAPING THE FUTURE OF WORK IN TECHNOLOGY

Introduction of technology in the organization facilitates the changes in the future of work. These are the following factors focusing on the future of work in the technology:

1. The propagation of troublesome technology is refurbishing to the business organization.
2. The new technology role has given impetus to the business operation and process. It has changed the business environment and evolved new workforce inclination such as gig, multiracial, and distinct worker. Technology has also changed the labor market to technological employee and virtual employee.

14.5 TECHNOLOGY AND EMPLOYMENT

It defined to the changes of technology on worker skills, job, and earnings. There are several research conducted which has proved that there will not be that much of changes in employment if new technology adopted in the organization. Although some said that changes will have impact on employment. However, for sustainability of the job for the long time period is dependent upon the required skills to work or to adopt the technology. In fact, productivity, and technology changes lead to growth in the earning. There are several factors are responsible to affects the level of employment due to change in the technology:

1. The time at what organization products and services are demanding into market;

2. The survival of new product in domestic and international market;
3. Redefining the process to produce maximum output with low labor requirement;
4. Effect of technology change on wages.

According to the Cambridge report, Inc. (1986), technology help to upgrade the required skill set in the employee that would be suitable for the work. Normally every employee must have two types of skills; Basic skills and Job related skills. Basic skills are one which acquire by the education system rather than organization. It involves knowledge about the job domain, reasoning, problem solving, and communication. Job related skills are one which is acquired by the organization needed to perform task. Basic skills help to workers to understand what they need to perform in their present and future employment. However, some study stated that job related skills can be acquired through on the job training.

Case Studies of Automation

There are several case studies has been analyzed the skills required for the adoption of information and computer technology. It has been stated that to adopt the new technology, skills needed to be change or learned new skills that match to the job requirement. If we talk about the change in the back office job, technology has eliminated the requirement of clerical employees although increase the demand of data entry employees. Many of studies stated that there is the requirement of increasing the scope of job through extending the range of its responsibility or by adding new task at same level. However data entry job would require the training about the computer related skills to fit into the organization job-changing environment. This case study stated that new technology in office automation helps to reduce the skills requirement. For example, office work can be reconstruct by the using the computers and high machine languages. Adoption of new technology divide the work into different workstation which would be required the different skills sets. For example, an insurance organization need to be stored and maintain different type of information of a customer for a policy such as; rating, underwriting, policy renewable and payment of premium, that would be required different employees with different skill sets.

14.5.1 FUTURE PROSPECTS FOR TECHNOLOGY AND EMPLOYMENT

Technology also has effects on earning of the employees. According to the U.S. survey, it has been stated that change in the technology gradually destroying the earning level and dominating to consolidating earning distribution. It is focus on the changes on wages level and defines the correlation between income distribution and technology change. It is found that in the service sectors income distribution is based on the occupational structure. Thus there could not be arise the changes in the distribution of income with no change in earning level as the spending structure has been changes in household people.

It has been stated that new technology can make employees superfluous, although it may be fictitious. In 19th century, British textile employees feared that they would be replaced as new technology was adopted into the organization. Due the inference of new technologies for employment, workers demolish the machines and burnout the house of John Kay, invented "flying shuttle." It is an assumption that the Adoption of new technology can replace the workers but would create substantial wealth also. However, it is also stated that weekly working hours would decreased by 15 Hours and monitory situation will improve also.

The new technology will also create the lucrative market, increased demand, stoke up the earning and new employment opportunity, results in resonate the new service opportunity. Thus it is sure that there would be annihilate some occupation. It is true that technology is reducing the task of the employee but it is giving the time to the employee to think about to upgrade their skills and for to make them more educated to become more desirable. What would be happen in future with technology change is completely dependent upon the organization policy and strategy as technology will also create some new job opportunity and these would be depend more on thinking, specialized, and analytical skills.

According to the bureau of labor statists, employment opportunity will be increased by 30% by 2020 in the computer field. In the human race we will be enter into fourth industrial revolution, i.e., 4.0 Industry is defining to the collaborating the physical, digital, and biological components.

14.5.2 BRUNT THE IMPACT OF TECHNOLOGY ON EMPLOYMENT

Figure 14.9 defining the relationship between technology, employment, and process. The ancient clincher argues that technological effects are transitory.

Therefore, the technology has specific impact on employment. The advancement and establishment of information technology (IT) is a contemporary illustration of the impact of new technology on employment. According to the EthanKapstein, Council on Foreign Relations has been stated that advancement in IT has promoted to the global competitions as well as opportunity. However, new technology contributing to the new job and growth prospects to the employees. It is also contributing to the economic development. Technological advancement focuses on higher productivity with limited resources.

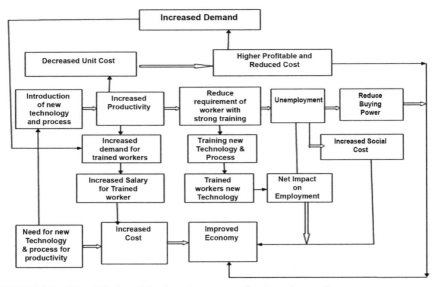

FIGURE 14.9 Simplified model of employment-technology interaction.
Source: Mario W. Cardullo (1997).

14.6 CONCLUSION

To manage the employee work, process, employment, and productivity organization are being lither. The organization using technology not to just do business and perform the task, for collaborate remote employee by adopting new concept of freelancer and gig employee. If organization is using automation then demand of new skilled employee must be there as they would require using the technology. Technology has divided the work in the task and all tasks can't be performed by the technology alone. Therefore, there is an employment for those employees who possess different skills set and acquired the training to perform the specific task. Technology has brought new fanciful future to the business organization where they can set their vision to complete into the competitive environment. Beside of reducing the job opportunity or task of the employees but it is also helping to upgrade their skills. The demand and supply of the products and services will be increased by using automated technology. It has good or bad impact on business organization. The impact of new technology on business is completely dependent on how to use it and the investment on the technology.

Impact of technology has strong connecting with the BS also. To accomplish their goals organization needs to adopt the technology. However, some organization are fails to identify the link between conceptual strategy and both technology and BS.

KEYWORDS

- business strategy
- employment
- employment-technology interaction
- productivity
- technology
- workplace

REFERENCES

Aeppel, T., (2015). Jobs and the clever robot: Experts rethink belief that tech lifts employment as machines take on skills once thought to be uniquely human. *The Wall Street Journal*. A1.

Archibald, R. B., (1996). *Working Paper-Measuring the Economic Benefits of Technology Transfer from a National Laboratory: A Primer.* Williamsburg, VA.

Floyd, C., (1997). *Managing Technology for Corporate Success*. Gower Publishing Limited: USA.

Hayes, R., et al., (2005). *Operation, Strategy, and Technology: Pursuing the Competitive Edge*. Wiley: United States of America.

Manyika, J., Chui, J., Bughin, J., Dobbs, R., Bisson, P., & Marrs, A., (2013). *Disruptive Technologies: Advances that will Transform Life, Business, and the Global Economy.* McKinsey Global Institute. Retrieved from https://www.mckinsey.com/~/media/McKinsey/

Business%20Functions/McKinsey%20Digital/Our%20Insights/Disruptive%20technologies/MGI_Disruptive_technologies_Full_report_May2013.pdf (accessed on 21 June 2020).

DeMeyer, D. A., (2008). Technology strategy and china's technology capacity building. *Journal of Technology Management in China, 3*(2), 137–153.

Robert, P., (2015). *Technology Road Mapping: Linking Technology Resources to Business Objectives.* University of Cambridge.

Ross, A., (2016). *The Industries of the Future.* New York: Simon & Schuster.

Vernet, M., & Mohammed, R. A., (1999). Linking business strategy to technology strategies: A prerequisite to the R&D determination. *International Journal Technology Management, 18*(3/4).

CHAPTER 15

Impact of Augmented Reality in Sales and Marketing

LALIT KUMAR SHARMA

Associate Professor, Jaipuria Institute of Management, Ghaziabad, Uttar Pradesh, India

ABSTRACT

Sales and marketing have been transformed into a new dynamics because of use of technological interventions. Techniques like augmented reality (AR) have bridged the gap between customer expectations and offers by the companies. In the current times, AR is adding new dimensions to the world by layering information onto experience through sensory experiences, visuals, and sounds. AR has combined the real world and virtual world to match the customer expectations efficiently. New and effective approaches have been developed by companies to understand their customers and serving them accordingly. Product promotion has been revolutionized because of digital marketing worldwide. Customer interaction is becoming the key to success with the use of AR. The components of digital world blend into perception of customers in real world and create sensation amongst them which are perceived natural. It does not simply display data. AR has moved way beyond science fiction. AR is gaining momentum in marketing research as companies are not very much aware about the perception of customers and their evaluation criteria for a brand. It is important for marketers to know about the changes in the brand attitude of customers with the use of AR.

This chapter is highlighting the importance of AR in sales and marketing and how it is used by the companies to create customer delight. Companies find it easy to engage the customers with their brands now. The primary objective of this chapter is to identify the importance of AR in the field of changing scenario of Sales as well as Marketing. The chapter reveals consumer inspiration and benefits drawn from AR apps and subsequent changes in brand attitude.

15.1 INTRODUCTION

Augmented reality (AR) is an innovative manner to present the information to the real world with the use of sound, video, graphics, and other sensors. There was a time when marketers were focused on content for marketing and used to believe that "Content is the King." Later, orientation has been changed as AR is providing interactive mechanism to customers to be both content creators and content consumers. Customers are interacting with offline and online products available for them because of AR technology. AR marketing contributes to a brand in development of complete brand interaction by:

1. Offering meaningful interaction;
2. Offering authentic communication;
3. Offering a gateway of brand engagement.

Today's experiential marketing landscape is generated with the use of AR and virtual reality (VR). If the use of AR and VR is made strategically by the companies, they can increase the brand penetration in market and generate more ROI. This will increase the customer engagement on social media and strengthen brand loyalty. If the customer engagement is more for a brand, chances of conversion are also high. This leads for more business ultimately.

Business value is offered by AR across the wide range of functions in Sales and Marketing. Benefits like cost reduction, opening new revenue streams, customer convenience, etc., are obtained with the use of AR. Easy implementation and continuous engagement is important for the success of AR technology. It creates the digital product experience to all users. AR can help the organization in the transformation as per changing dynamics in the global market. Companies are getting fast returns on investments made in AR creation. AR is gaining the space in our daily life and providing desired experiences to us. Some basic activities of our life like newspaper reading, shopping will not remain same due to technology interventions. AR was primarily used for branding in public domains up to year 2014 where typical use of cameras using persons and their surroundings was made and virtual elements were added and displayed on screen. It helped in creating awareness amongst consumers for AR as a marketing tool. The exposure of AR tool was enhanced with the introduction of Snap Lenses in year 2015. Decoration of 30 million snaps was done with lenses by advertisers in first year. The brands have achieved scale with AR first time with high reliability factor.

15.2 AUGMENTED REALITY (AR) IN MARKETING

Marketing has gone through with so many transformations from the era of production concept to societal concept and now digitalization in market. Orientation of marketing variables has been changed in this journey. As we know that Marketing is the process of:

- Identifying the basic needs of customer;
- Developing the product to satisfy the needs;
- Devising the strategy how product will reach to customers.

All the steps in marketing can be taken by marketers effectively with the use of AR as it helps the companies to gain the insights of their customers and making their offers as per the choice of customers. AR is the combination of virtual products and real world and the people get the chance to interact with these products in real time as per the convenience of them. The computer vision of integration of Graphic, Video, Sound, and censors with the use of camera makes an excited AR technology which combines the virtual products and potential users. Organizations provide useful information to the customers in real world with the use of AR and generate momentum in the business by catching the eyes towards the brands offered by them. AR provides the feelings to the customers due to interaction between real world and virtual world. AR is conceptualized with the objective of supplementing the customers in real world, rather than focusing on creating artificial environment only.

AR is helpful to the marketers in development of marketing strategies. Marketing companies design the multiple strategies to assess the value of their products and market first and later, they can design the brand engagement with the use of AR. Information about target audience is very helpful in combining AR and VR. Use of AR technology helps in determining the effectiveness of any marketing communication strategy and developing brand voice. Customer experience can be enhanced because of an informative and interactive utilization of AR marketing. It is adding value to customers in a product. By increasing the value index or benefits (functional and emotional) to the customers, customer delight becomes easy for the company. Users can get additional information about products in a store and can scan about prices or nearby availability of stores. Live AR technology is helping the marketers in many industries like Hotel, Education, IT, Gaming, etc. Companies in the business of gaming are continuously deploying AR technology for generating more entertainment to the customers. Live AR has created wonders in retail sector.

AR is giving another important tool to marketers for generating more sales and enhancing brand value. Markets are very much driven by needs, wants, demand, and desire of customers. With the advancement in technology, desires can be changed and important for the organization to run with the pace of change. Customers like to interact with brands before making the decision to purchase now. AR is very helpful in creating experience for the customers through mobile devices.

15.3 WHY TO USE AR IN MARKETING?

Bringing innovations in marketing strategies is always a challenge to the marketers. They need to focus on image of the brands as well as competitors. Integration of various factors is required as Marketing is widely spread in terms of domain. Fields like Sales, Distribution, Integrated Marketing Communication, Consumer Behavior, Marketing Research, and Branding are integrated for the success of a marketing strategy. Use of AR in marketing is becoming important with the new challenges in business day by day. Every marketer wants to get success by delivering best products and services to their customers. As we know that the likings of the customers are changing too fast and it becomes difficult for the marketers to acknowledge these changes all the times. Therefore, working on the successful marketing strategies is not the easy task. The big challenge remains for the marketers to know about the changes in the daily course of action. The use of technology like AR is helping the companies to understand their customers in a better way as it is providing an interactive mechanism to be in touch with them. AR helps the marketers in creating value to their customers. Basically a customer looks for functional value as well as emotional value. Emotional values can be created easily with the use of AR. There are several reasons for the companies to use AR in marketing for better performance of business. Some of the important reasons are discussed below.

15.3.1 BUZZ

The use of AR in business is very new to many markets. Therefore, it may create the buzz and excitement to most people to see and experience something first time in life. This can help in motivating the customers to purchase even when they are not planning to buy. New brand experiences are always a motivation factor behind the purchase by customers. AR can do this job very effectively. The companies can sensitize their customers by creating buzz even before launching the product in market.

In a real-life example, we can see Nintendo's Pokemon franchise. Most of the customers had not seen a well-done AR video before Pokemon Go. Pokémon Go's revenue $207 million was the highest ever of any mobile game in first month of launch. The main reason of success was effective storytelling combined with new AR technology.

AR is used by marketers as part of an indirect sales and marketing strategy. Through which marketers try to popularize their brands. Companies try to create a novel and fun making AR experience that results in creating a lot of buzz about the brand. Companies need to ensure a careful execution of AR to achieve optimum results. New experiences always create memories if they match with the customer's mindset for a brand and it can be done by companies with the right use of AR. Customers always love to buy the brand which makes real promises and keep them satisfied. Goodwill can be generated with the use of AR to influence the buying decision of customers for a brand. Companies like Pepsi, Uber have used AR very intelligently to influence the customers. Pepsi Max Unbelievable Bus Shelter was similar experience to give unexpected moments to commuters. This helped the company to create discussion about them.

15.3.2 GAMIFICATION

There is a positive correlation between customer engagement and sales revenue. It means that when customer engagement is high, sales is also. When customer is engaged with the brand, it leads for high retention and chances are high for a brand to reach on the top of the mind of a customer. Customer engagement is not only important for new customers but also for existing customers. A study by Gallup reveals that the brands which have successfully engaged their customers found 63% lesser customer attrition and 55% higher share of wallet.

The customer engagement can be increased with the use of Gamification as per latest trends. With the use of AR, more time is spent by potential customers in interaction with marketer's app or promotional campaign. It may be looking at product's 3D model or watching a movie trailer which appears on pointing the screen.

15.3.3 CUSTOMER CONVENIENCE

Understanding or experience of the brand generated with the help of AR marketing ensures the convenience to customers in deciding the products to purchase. Innovative use of AR technology creates the platform for all marketers to showcase their brands to the potential consumers and influence their choices and make a smooth process to decide the brands to their customers. All potential customers always want to have trial before purchasing them. Many traditional concepts like automobile test drive, sampling of cosmetics, etc., are proving the success of sales strategy. AR is creating augmented shopping experience to users in current retail industry. For example, Shoppers Stop is using magic mirror in some of its retail outlet for selection of apparels by its customers. Dressing rooms are always inconvenient for customers for trial of fitting of clothes. Customers need to carry stacks of clothing items to dressing room and employees are required to take care of discarded items. AR application helps the customers to select the brand without wearing it. It creates the facility to the customers to try number of samples out of hundreds with ease and select the best as per their requirement. It helps the companies in storing proper inventory also. AR applications are increasing in business as per the benefits presented to companies and customers. Digital samples of the brands help the customers to make wider choices for their needs. Companies are making use of social media also for better connectivity to their target customers with the use of AR. These digital samples are developing customer convenience by providing knowledge, interactive medium for sharing and resolving doubts (Figure 15.1).

FIGURE 15.1 Customer convenience through AR.

Many cosmetic brands like L'Oreal, Sephora, and Perfect Corp. are working in partnership to provide experience to their customers to see how they will look in makeup digitally. AR is helping in creating online sales strategies as customer is always judging any makeup item by modeling it on themselves.

15.3.4 ASSISTANCE

Digital platforms are gaining importance in business to market the brands by marketers. Brands which fail to generate the registration in the mind of customer are never considered for the purchase. Marketers always try to create an image for their brands against the competitors so that they are the preferred choice at the time of buying. Digital platforms help the companies in creating easy availability of products and providing product-related information to the customers. A digital component is added to a business above the products and physical locations to assist the customers with variety of products offered by the organization. Customers are able to scan the availability of the products in a retail outlet if the use of AR has been made by a retail company. Customers can gather all sorts of relevant information with AR and without the help of employees. AR technology keeps on providing assistance to customers as per choice. It assists in developing brand experience to customers and even mood elevation is done many times. Users can scan the objects in a Starbucks coffee shop and access a virtual tour. High use of AR in entertainment industry is helping in boosting up the business and attracting the clients. Within the paint industry, Asian Paints, Nerolac, Berger paints have adopted AR applications to help their customers to feel the impression of colors and develop their choice of colors.

15.3.5 BRANDING MATERIALS

Though AR is an emerging technology yet the investment by Apple in ARKit and Google in ARCore clearly shows that AR market is going to be there in future. Branding materials like brochures can be transformed into different level by adding virtual component with the support of AR. Customers are scanning the printed materials and accessing the multiple features with more information with their Smartphone. Dynamic virtual presentations of a brand create more impact on customer's choice. It leverages the experiential learning of brands to the customers. Even AR can be used in a business card that allows the users to get in touch with the brand through a phone call, email, or LinkedIn. AR helps in enhancing the brand engagement with virtual branding materials.

Branding material is supporting the marketing communication strategies and helping in brand registration in the mind of customers. Few examples of using AR in making print communication more effective are:

1. **Lead Generation:** AR is playing an important role in generating leads for the business. Many companies are using AR for continuous interaction with the customers and trying to create positive impact about their brands in the mind of customers. This process is sensitizing the customers towards the brand and helps in generation of unique selling proposition (USP).
2. **Knowledge with AR:** The major objectives of the integrated marketing communication are creating awareness, persuasion, and reminder to the customers. All three objectives can be fulfilled easily if the marketers are making effective utilization of AR. As AR is giving the chance to the customers to interact about the product, so the desired information can be given to the customers for knowledge purpose.
3. **Cost Reduction:** The cost of promotion can be reduced with the use of AR as compared to conventional printing branding material. With the use of mobile app, AR is offering better connectivity with customers with lesser cost. Moreover, interactive communication is generated by marketers which ensure the better impact of communication efforts. Money on postage is saved.
4. **Incentives with Augmented Reality (AR):** Marketers can use AR component on their print piece and offer the incentive to customers for interacting with their digital channels of brands. Incentives like coupons can ensure high engagement of the customers. Augmented mails allow the customers to go through with personalized promotion and encourage them to interact with the brand. This leads for customer conversion as outcome.
5. **Delight Generation:** As we know that when benefits are more than the cost paid by customer, lead for customer delight. With the use of AR, marketers try to enhance the value index of their brand. AR is very much helpful in enhancing emotional quotient of a brand. Impulsive buying is very high for daily life products.

15.3.6 TRANSFORMING B2B EXPERIENCE

There was a time when use of AR was confined to B2C segment mainly and need was not felt in B2B segment. It was considered that B2B is a logical business process so personal selling is an effective mechanism to deal with the customers. Now, the need of changing the B2B customer and vendor experience has been realized and AR is making significant contribution in B2B Sales. It is always crucial for sales force to deal with customer expectations and limitations of offer provided by the vendor. AR has high potential to improve the complete sales process. Importance of sales presentation material is very high in B2B sales. AR can help in making dynamic presentations for the products. The use of digital devices supports the salespeople by offering a 360° look of their products to the customers. This creates the real feel of products. Old methods of presentations like brochures, flyers, and PowerPoint presentations (to a certain extent) have been replaced with the digital devices now.

AR sales tools provide the detailed information of the products and give them access to customers to interact with the product in detailed manner as per desire, from the different prospects of overview. Bringing the product and information in discussion room, AR allows the clients to have access to information required for decision-making. AR is a great support when companies have to deal with customization options. In general, specifically tailored products are desired by many customers. Such demands are not easy to handle by vendors all the times. AR tools give the freedom to customers to design their own products and it gives better understanding to companies about the needs of their clients. The inputs from customers can be digitally transformed in real time to the companies and a feedback mechanism can be developed in association with customers so that they can get same product what they exactly want.

AR is not only supporting a business in pre sales or concurrent sales but also in post sales process and support. Many times, customers are looking for user manuals or online knowledge data for resolving their queries but they find it difficult to navigate. This can result in dissatisfaction to customers and they can start searching for some other vendor for business purpose.

15.4 THE ADVERTISING INDUSTRY

Advertising is always instrumental in deciding the success of brands for the companies globally. Conventional advertising approaches are losing importance gradually. AR is helping the organizations to develop unconventional but more effective mechanisms to promote the brands and engaging the customers. Smartphone and other mobile technologies have taken place in daily life of the customers and internet has become the major source of information to the customers. AR helps in providing an innovative media format that integrates virtual information to the perception of users in real world. AR ads are helping the marketers in connected with the customers emotionally. Customers get the feeling of playing with a video game with AR ads. It helps in building emotional connect with their customers and motivating them to buy the product. AR creates a new method of reaching to the customers. Many companies have started using AR technology to reach their customers. Those companies which are not sure about the use of AR ads and its worth should understand the benefits of AR ads first and understand the utility of technology in daily life of people. This will help them to grow more in the business.

15.4.1 REASONS TO USE AUGMENTED REALITY IN ADVERTISING

Digitalization is growing in current market everywhere. The number of internet user is increasing year by year. It generates a lot of work for marketers and advertisers. It is expected that spending on digital ads will grow from $283 billion in 2018 to $517 billion in 2023 (Figure 15.2).

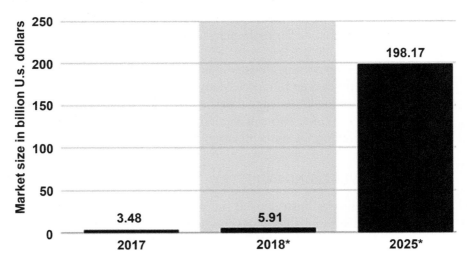

FIGURE 15.2 Augmented reality (AR) market size worldwide in 2017, 2018, and 2025 (in billion U.S dollars).

Source: Statista.

AR ads are used mostly by early adopters but they getting popular as people using AR are increasing. As per Statista, it is forecasted that the market size of AR will grow from 3.48 billion in 2017 to $198.17 billion in 2025. There are many reasons for the growth of AR ads in future:

- Emotional connect;
- Economical advertising;
- Increasing sales;
- Interactive communication.

15.5 DIGITALIZATION IN GLOBAL MARKET

Digitalization is making its mark in global business scenario. All companies are trying to create digital footprints for themselves in a tough competitive market. Social media is playing a big role in the success of digital era. As per Digital 2019 Q4 Global Digital Statshot report which was developed in association with Hootsuite, many important observations have been noticed. Some highlights are as given below:

1. **Big Increase in Numbers of Internet User:** The number of internet users is increasing worldwide with annual growth of 10% in last one year and approaching to 4.5 billion very soon. Many impressive achievements will be seen in 2020 as per current trends.
2. **Increase in Social Media Users:** Because of availability of new data, social media users are increasing every year. This increment is not due to increase of numbers in new users only. India is one big reason for this growth as WhatsApp user number has been crossed to 400 million active users. It indicates that almost 30% of Indian population is using WhatsApp every month. This number is 100 million more than Facebook active user. A lot more opportunity is there for the growth of social media in India. India is a one of the biggest social media market. Other platforms like Facebook and Tencent are also contributing in the growth of social media user numbers, but interesting trends have been emerged at the level of individual platforms, as it can be seen from following details:
    - **Slowdown in the growth of Facebook's Ad:** Slow growth has been registered in the numbers of advertising audience of Facebook's core product over the past three months from July to September 2019. Only 8 million new users have been added in this duration. The growth rate in third quarter is only 0.4%

which is lesser than the 3% growth registered in second quarter of the year. When we see the figures of individual countries, story seems to be more interesting. It is suggested by Facebook's own tool that reach of advertisers has been reduced in US and Indonesia in July-September 2019 quarter compared to last quarter. It indicates that advertising market may be shrinking in these two markets. Though Facebook still have the largest youngest audience among all social media platforms worldwide, latest numbers of the company shows that young advertising audience is also shrinking. As per the reports of Facebook's advertising tool, advertisers have reached 2.6% lesser users in age group 13 to 17 in third quarter as compared to second quarter 2019, and 2% lesser users in the age group of 18 to 24. There can be many reasons for the decline in numbers as if Facebook has removed many accounts who have breached the terms and conditions.
- **Pinterest:** As per first earning report of Pinterest, platform has 300 million active users in a month across the world and showing the annual growth of 30% compared to second quarter of year 2018. Some interesting facts about Pinterest are advertising audience shows that woman is contributing significantly in the share of total audience. In US, advertisers on Pinterest can reach same numbers of women of age over 30 years as on Instagram. Interesting fact is that total global advertising audience of Instagram is almost 6 times bigger than Pinterest.
3. **Rise of Asia:** As per App Annie's latest data, 6 Out of 10 most used mobile apps in world are owned by Chinese companies (Rest 4 is owned by Facebook). Mobile game market is also dominated by Chinese organizations and almost contributes to 40% of all app downloads globally. This confirms the dominance of Chinese companies. Not only Chinese organization but also organizations from Korea and Japan are also delivering many popular apps.

15.6 MARKETER'S ROLE WITH THE USE OF AUGMENTED REALITY (AR)

AR has instrumental effect in the role of a marketer. Marketers are able to design highly innovative brand communication with the use of AR and they ensure the customer engagement simultaneously. Now the scope of traditional one-sided communication is declining and marketers are bringing more innovative communication approaches to influence the customers. All marketers will have to focus on how effectively they can use digital platforms to convey their brands and how effective they are in developing interactive mechanism to deliver customer satisfaction. Marketers should understand the similarities and differences between AR and other digital platforms to leverage more opportunities for marketing campaigns.

As the number of persons using Smartphone are increasing everywhere, so marketers should consider it as an opportunity to get connected with them and display their brands with interactive communication. Many customers are ready to use mobile apps for their convenience; this is an opportunity for marketers. Therefore, marketers should consider all the devices used by consumers for leveraging easy adoption. An app with the inclusion of AR technology will be less difficult to use as technology is embedded in the application.

15.7 STARTUPS AND AUGMENTED REALITY (AR)

India is growing very fast in business in terms of number of startups every year. One significant parameter for the startups is the use of type of technology in business. Technology is the base for most of the startups now. Startups like Oyo, Ola, Paytm, etc., are making business impact outside the country also and considered very high for their business innovations. One major reason of success for these brands is effective utilization of AR so that customers are finding such companies very close to their needs and their readiness to use the brands is very high. Some benefits offered by AR to such startups are:

- Better connect;
- Transparency in dealing;
- Effective communication;
- Customer convenience;
- Mass availability;
- Customer engagement;
- Trust;
- Wider acceptance.

As mostly startups are funded by external agency or partner, it is vital for them to experiment newer things in marketing so that customer interest is maintained in future and investors can have better returns. Startups keep on experimenting with AR in order to facilitate their customers.

15.8 PROJECTIONS OF AUGMENTED REALITY (AR) IN MARKETING

1. **Better Client Conversion:** It is always a challenge for the marketers to attract, convert, and retain the customers in the era of cutthroat competition. Experiential marketing is a method adopted by many organizations to attract, convert, and retain the customers. AR and VR will assist the companies in understanding more about the customer expectations and how to delight them. Visualization and experience creates value to the marketing efforts and help in gaining more business.
2. **Content as Driver:** One big reason of AR and VR is content availability and reliability. This will keep on happening in future also. People will be looking for more sources in the search of quality and authentic contents. To offer more value to customers, content requirement will be very high on digital platforms. Companies in the business of content development will be more in numbers and provide assistance to marketing.
3. **Creating Better Customer Experience:** Advertising with the use of AR and Virtual Reality will be used by companies more to narrate the stories of brands. This will lead for more customer experiences for a brand. Creation of interactive communication and experience around the brand content will create the desired responses to marketers. Conventional advertising practices are very limited in terms of customer interaction so the effectiveness of such advertising is not very high now. Therefore the impact of experiential marketing is going to be very high and all marketers have to develop new approaches for creating better customer experience and engagement. Experiential marketing will be successful only when it is bringing digital reality into real world.
4. **High Degree of Customization in Market:** Advertising will be more integrated in our daily lives with the help of AR. Marketing and advertising will be using more real-time data based on the preferences of us. Enhancing the brand engagement and experience will be the ultimate goal of all marketing initiatives.
5. **Exploration of New Products:** Involvement of AR and VR will allow consumers to have new experiences when they explore new products or services. At the time of exploration of new products and services, many new thoughts will be developed in the mind of customers and companies would like to map these thoughts. Companies can differentiate their brands on the basis of these thoughts if

they are able to map them accurately. Customers will be able to have a 360° view of a product before they decide to purchase.
6. **Bigger Investment in Technology:** AR generates live experience with the involvement of digital elements. Marketers will have to invest in technology to ensure innovations in their brands. Future will be highly technology-driven so the companies are required to either replace the existing technologies or upgrade them. Those companies which got failed in adoption or development of technology will not be able to survive.

15.9 CONCLUSION

AR is combining the virtual world and real world. It is important for the marketers to ensure the use of AR and VR in their marketing practices for attracting, converting, and retaining the customers. The use of AR will enhance the business prospects to all marketers. AR is helping the organizations in assessing the preferences of their customers and develops the brand communication accordingly. AR is not merely the tool for promoting the brands but also helps the organizations to decide their sales and marketing strategies. AR will upgrade the marketing approaches to get better insights of the customers with the development of continuous interactive communication. Benefits like Buzz for brands, Customer convenience, Gamification, Branding materials are achieved due to the use of AR by marketers. Startups are getting the benefits of Better Connect, Transparency in Dealing, Effective Communication, Customer Convenience, Mass Availability, Customer Engagement, Trust, and Wider Acceptance because of the use of AR in their marketing practices. Organizations can upgrade their existing technology in order to get maximum advantage of AR in the business. Content is an important factor for the success of AR in marketing. Content should be designed as per the taste and preferences of the customers in the market. Customer experience of a brand is important for the decision to purchase the brand. Technology is not only helping the people to view virtual images in the real world but also creating the interaction of customers with these images and objects. All these things lead for improving customer experience with the use of a camera on their device by a marketer. The sales cycle can be shortened with the use of AR and cost of marketing can be reduced. All types of businesses like B2C, B2B, and C2B can have more advantages and deliver better customer delight with the use of AR. Apart from the US and many European countries, Asian countries like India, China, Korea, and Japan, etc., are also contributing a lot in the growth of digitalization.

KEYWORDS

- augmented reality
- brand engagement
- digital media
- digitalization
- marketing
- sales
- social media
- virtual reality

REFERENCES

Alex, O., (2009). *Adapted from: Olwal, A. Unobtrusive Augmentation of Physical Environments: Interaction Techniques, Spatial Displays, and Ubiquitous Sensing.* Doctoral Thesis, KTH, Department of numerical analysis and computer science, Trita-CSC-A, ISSN 1653-5723; 2009:09.

Andres, H., & Mark, O., (2004). MUM 2004 October 27–29 2004 College Park, Maryland, USA Copyright 2004 ACM 1-58113-981-0/04/10. $5.00.

Azuma, R. T., (1997). *A Survey of Augmented Reality Presence: Teleoperators and Virtual Environments, 6*(4), 355–385.

Azuma, R., Billinghurst, M., Schmalstieg, D., & Hirokazu, K., (2004). *Developing Augmented Reality Applications.* ACM SIGGRAPH 2004 Course Notes.

Bulearca, M., & Tamarjan, D., (2010). Augmented reality: A sustainable marketing tool? *Global Business and Management Research. 2, 3,* 237.

Dhiraj, A., & Sharvari, G., (2015). *International Journal on Computational Sciences and Applications (IJCSA), 5*(1).

Fenn, J., (2010). *Executive Advisory: Six New Technologies That Will Reshape Your Business.* Stamford: Gartner.

Kengne, P. A., (2014). Mobile augmented reality-supporting marketing. Using mobile's augmented reality-bases marketing applications to promote products or services to end customer. *Program in Business information Technology,* Autumn-2014.

López, H., Navarro, A., & Relaño, J., (2010). An analysis of augmented reality systems. *2010 Fifth International Multi-Conference on Computing in the Global Information Technology (ICCGI 2010)* (pp. 245–250). Valencia.

Olivia, Y., (2014). Augmented Reality, Augmented Reality Application, Augmented Reality.

Owyang, J., (2010). *Disruptive Technologies: The New Reality will be Augmented.* Customer Relationship Management Magazine.

Pfeiffer, M., & Zinnbauer, M., (2010). Can old media enhance new media? *Journal of Advertising Research,* 43–49.

CHAPTER 16

Theoretical Perspective of Role of Technology on Business Environment

JASMINE MARIAPPAN,[1] CHITRA KRISHNAN,[2]
KARTHICK SHANKARALINGAM,[3] and SYED MOHD. ABBAS[4]

[1]Ibra College of Technology, Oman, Muscat

[2]AIBS, Amity University, Noida, Uttar Pradesh, India

[3]Freelancer, Oman, Muscat

[4]RSM, Raymond Apparel Ltd., Mumbai, Maharashtra, India

ABSTRACT

Information technology (IT) describes any technology that is necessary for business performance to generate, process, and disseminate information. For the business sector, IT is an important management tool in automating efficient information processing. No matter how large your company is, there are tangible and real technological advantages which help make money and bring results that your customers need.

We definitely live in a world in which technology forms the future. There's always been company since man's early days. Although it started with the basic trading method, business would not be the same without technological progress as it is now. In the advent of technology, company principles and models have been revolutionized. Technology has created a new and enhanced way to do things. This offered for business transactions to be done quicker, easier, and more efficient. Technology has allowed the global market to reach a greater extent. A fundamental example is the Internet, now a common marketing tool to encourage more consumers to use the products and services of different companies. This study takes into consideration valuable theoretical work to examine the role of technology in business world. In order to understand better what these technologies are able to do, will demonstrate what a few different firms are doing to transform business as we know it.

16.1 INTRODUCTION

Information technology (IT) describes any technology that is necessary for business performance to generate, process, and disseminate information. For the business sector, IT is an important management tool in automating efficient information processing. No matter how large your company is the technical advantages that it's tangible and true to help you make money and produce the results that your clients need.

The profitability of the business world has been greatly increased by technological advances over the past few decades. Apps, computers, and the Internet have been used by companies to turn their companies from local businesses to competitive on the national and global markets. Most businesses have adapted to this by automating their business processes and gathering and taking advantage of information relating to industry.

Tech also has made businesses versatile and adapts their operations to new and improved technological developments. Several innovations, as either advertisers or consumers, are used by us. They wouldn't be wrong in either case to say that technology is powered in the world today. At first, businesses were entirely labor-based. However, no company wants to lag behind with the growth of technology.

Many development activities in business include accounting systems, IT systems, pricing points, and other easier or more complex devices. Even the machine is a technical product. It is not understandable indeed to invoke the notion that we should go back to the days when everything has been done manually.

The productivity of the business world has greatly enhanced technological advances in recent decades. Companies have been turning their businesses from local companies into national and global competitions using software, computers, and the Internet. Many companies responded by automating their business processes, collecting, and using information related to industry. Technology also forced companies to remain flexible and adapt business to new and better developments in technology. Business owners used to have very nothing tool: little more than simple machine and hardcopy records at their fingertips.

Business owners today are much more capable than their predecessors of meeting their obligations by using a variety of technical resources. Through the use of these devices, companies, and workers have a number of business advantages. We recognize that the business sector generates profit-making products and services. IT defines any technology that is essential for business performance and is used to develop, process, and disseminate information.

In order to monitor information for the business sector's profit growth, IT is essential for producing products and services. The number of workers needed by the use of information technologies is usually lowered by changes in automation. The total cost of producing products and services for companies limits the amount of IT usage savings. This affects the financial results of an organization significantly positively.

16.2 WHY IS TECHNOLOGY IN BUSINESS IMPORTANT?

Technology helps businesses, without software, to do more, better or quicker than you can. Different industries and firms rely on technology in various ways, but the technology benefits of enterprise include enhanced communications, automated development, stock management and keeping of financial records. Some of the following was explained:

- Innovation extends the range and growth, demonstrating the social and industrial value of the innovation, of a vast number of internal and external contacts. The technology mainly affects the ability of a business to connect with consumers. It is important to communicate quickly and clearly with customers in today's busy business environment. After hours of your inquiries, consumers can find answers to websites. Fast transport options allow companies to carry goods across a wide geographical area. As consumers use technology to connect with an organization, companies have a greater public image that results in better interactions.
- Technology also helps an organization to address its cash flow needs and retains valuable resources such as space and time. The inventory technologies of the warehouse help business owners to understand how best to control the cost of maintaining a product. Management can save time and money by meetings on the Internet and not through the use of appropriate technology in its corporate headquarters.
- The automated processes that technology can bring are that in efficiency. The explanation for this is the limited resources expended on corporative output to produce quality products and provide faster service to more customers. Data is also stored easily and

integrally. This decreases access to information that is confidential and important. The data can also be obtained and analyzed easily for trends and future changes that are important to decision-making processes.
- Building on the dynamic process network, business requires contact, transport, and more fields. Technologies in other fields have only further guided industry. The advances of technology have contributed to globalization. Everybody can do business anywhere in the four corners of his house.
- Companies should have enough inventories to meet demand without spending more in inventory management than they need. Inventory management systems monitor the amount of each object in a company that triggers an additional order if the amounts fall below the given level. The inventory management system is the most common use of these techniques.
- In the long term, technology will save companies money. Although the system installation and alteration will not at first seem like cost savings, it appears later. No matter what you use, you will find exactly what you need. This ensures that you save money by using technological or apps instead of paying others to do so or actually trying to create your own network. You can always use technology to slash the costs, and that is just what businesses do every day.
- There are rapidly deteriorating days in large rooms, filing rows and the mailing of papers. Today, most organizations are storing digital paper versions on servers or storage devices. Such records are open to all members of the company without hesitation, regardless of their location. Organizations are economically able to store and retain enormous amounts of historical data and workers have immediate access to documents.
- One of the greatest advantages of using technology is that you can begin automating activities that have taken a long time. Instead of spending some time on small tasks, you can use technology to make such tasks easier for you and your workers. You will not have to try almost as much to make sure it is done correctly and in good time.
- Many businesses that spend more in marketing and publishing today in the same sector. We will keep the achievement up to date. No one wants to work with a barely noticeable online company. For estimating and trading your rivals using algorithms. Technology uses effective management tools for online sales.

- New technologies will be used by companies from an external perspective to develop and enter new markets. Digitally advertise businesses in need of forward-looking traffic. Since the company is a brick and mortar business, the marketing strategy will include technology if you plan to increase your profits. Remember, a highly experienced IT team is the secret to your success.
- Employees expect their managers to perform their tasks efficiently with the new and highest technology. It is harder to compete against companies that completely use technological progress when your own business is lagging behind in their inventions.

16.3 CHALLENGES IN TECHNOLOGY FOR BUSINESSES

To business owners looking to simplify their business, increase productivity, and make life easier, technology can become a very valuable tool. Good IT use can help companies tremendously, but poor use of technology will lead to more customers' failure to deliver unreliable systems. There are a number of obstacles for business owners today when it comes to management of technology needs (Ray Ramon, 2007). Companies are affected by changes in technological environment. There's been a mistake.

Technology is difficult to introduce in a complex business world and involves a variety of management instruments and processes to take account of changes occurring (Phaal et al., 2005). First, the dedication to technical plans from top management is important, as Haywood (1990) suggests. Second, it is also two challenges for organizations to recognize the appropriate technology and pick its source. In addition, individuals within the organization should be able to use new technology and invest in increasing value. In such a way that technological change and development are possible, organizations should be coordinated and structured. In order to become more competitively and achieve their short-term and long-term goals, firms should also enjoy this change.

Technological change brings challenges, opportunities, and risks to organizations. Most businesses can develop their products and processes with technological changes and can even create new products and processes to expand sales and profits. The challenges include a lack of technological awareness, an inability to prioritize which innovations are of the greatest importance and a specific time span and how technology can be incorporated into an organization and secured. Some of the technological challenges faced by companies are as follows:

- Safety is of the utmost importance for companies today. You can also sell your website credentials, so nobody can hack you on a laptop today. Smaller businesses often have goals as they offer bank information and customer lists to rivals, terrorists, etc. The website also provides authentication and monitoring of the network. This allows you to protect your company from threats calmly and comfortably. It is important for you to understand technology if you want to grow your business today. It moves the world at velocities like any time in history. You face technical challenges in order to connect more efficiently, sell more, and develop every aspect of your business, so that you can take up the market accordingly.
- Controlling is a task that is challenging but important. You always have to up-to-date the apps. If you don't, you can miss key functions that make your operations. This can be done in many different ways, including in accounting, marketing, and administration. Make sure that all systems are running your devices and your updates are patched.
- The main technological roles for companies worldwide are drastically lacking. One of the main shortages is seasoned industry leaders. A company cannot only employ more "technologies" for success, but needs knowledgeable, experienced technology leaders. In addition, how are you doing these roles better and getting the best talents? In this blog series, I'll include some suggestions.
- There is still a growing rate of change. Your clients expect more, their competitors move faster and their needs change quickly. During the years, major projects in your technology organization were planned and carried out, but now must be completed within days.
- How will the company use digital technology to improve companies and give consumers higher value in all facets of the organization? The digital interface is a major factor in the experience with consumers and can inform the entire organization. Digital transformation often occurs in a rapidly changing environment that improves the efficiency, competitiveness, and productivity of a company. The award passes first and some very fast supporters, leaving them to catches up and fight for survival.

16.4 NEW APPLICATIONS AND INNOVATIONS IN TECHNOLOGY

It is no surprise that the Fourth Industrial Revolution and its signature new technologies will drive business growth, work creation, and demand for skills. Large companies are investing heavily in new technologies. Most IT and data processing, marketing, and consumer segmentation instruments are held. Others depend on payroll systems, onboard workers and other time-consuming activities. For daily activities, some businesses integrate virtual reality (VR).

Which technology are you expecting to have the most disruptive impact on businesses over the coming year, with many new applications and inventions involved? The following are the best answers:

- **5G Networks:** Already telecom companies have started building 5 G networks. 5 G provides quicker communication and reduced power usage at lower latencies. It means that mobile access is cheaper than most wireless cable in the United States. I assume it will promote creativity in the web of things, in intelligent homes, in increased reality in the virtual world, in medical care, media, in cloud services, etc.
- **Mobile Applications:** Today, people are able to search and read about businesses. Mobile technology and everything are now connected to people's worldviews. It involves business, as businesses are now designing mobile apps to reach their target customers quickly if they need them. Today, without a mobile app, it's difficult to find a business and some businesses want to find ways of using this technological aspect even more.
- **Advances in AI:** AI has been used for improving efficiency, productivity, and outcomes in the manufacturing, science, medical, and other industries already. The 2017 AI global executive study and Research Program MIT Sloan management review found that 85% of management believes AI will be able to gain or maintain their company's competitive advantage.
- **Increasing New Technology Adoption:** Technological innovations are related to the top five market competitive drivers by 2022. It's no wonder that the Fourth Industrial Revolution and its new technologies will stimulate entrepreneurial growth, employment, and demand for skills.
- **Prescriptive Analytics:** Essential analytics were a saintly grail from the beginning of proliferation of data analysis goods. So far, the most focus of analytical tools has been on descriptive as well as predictive applications. It allows for better organization and representation of historical information so as to provide a better understanding of the changes which have occurred and to use it to forecast events and actions in the future. Prescriptive research has, however, already been implemented in the energy, gas, and health industries.

- **Cloud Computing:** You probably heard about cloud computing and how this could impact your company, even if other technological advances have not been taken into consideration. This is not just a way for you to put the whole digital archive online, but it is also a way to further develop. Mobile computing is connected to mobile applications, and businesses do not have to worry about being able to satisfy customer demand or missing their servers or Internet during large-scale smartphone sales. The right service, for which businesses use the cloud, will discard concerns of crashes and loss of information.
- **Decline of Application's:** The next challenge for businesses that are worried about the money they have made in software and the competition of the different app stores is to find and update applications. Most people believe it might be more convenient and cost-effective to make them available from the cloud or to scrap them all together in favor of revolutionary Web Apps.

16.5   EXAMPLES OF TYPES OF TECHNOLOGY THAT BUSINESSES CAN USE

There are many forms of technology suitable for small companies and can help any business go further. Whether you are just starting or are just looking for a way to improve the technology of your company without spending too much money, these companies will help you to reach your goals. These are only a few businesses that have an impact on their advancements in technology. It is amazing that new technologies are created by these different companies that are able to change radically how business is done.

Industries have tried and improved technology over the years. New types of technology, from the stenotype to postal, have increasingly enabled companies to succeed. Software allows us to work even more effectively and productively before now, which eventually changes the way we work. Workplace technology went beyond email or other basic tasks with the device of a desktop or laptop.

1. **Guideline:** Many small companies may find it difficult to explain different benefits for workers. If you are struggling to find an affordable way to provide your workers with a choice of 401(k), then it is time to consider guidance. For the employer and the

employee, the Guideline makes the 401(k)-procedure easy. For a corporation, it is easy for its workers to create a 401(k) option and it is even easier for employees to register and start saving their own pension money. As you say, in a retirement account and more, it's all you want.

2. **Gusto:** HR is a major component of any organization and if you want to make sure that your employees are satisfied, some sort of HR practices need to be developed. There's so much about HR, and if you haven't got the money to hire someone to take care of this full-time, it might be Gusto. Gusto offers you a way of handling HRs, benefits, and even payrolls via its platform. You have the option to choose only one of your products or the whole product range for your client.

IBM, Intel, Kinaxis, Zoom, Zapier, and a number of other business-to-business examples are corporate leveraging technology. For example, IBM's B2B Collaboration allows companies to save and exchange data in a secure environment. Users can migrate files on a single platform, transform the data to valuable insights, and monitor their performance and behavior. With these statistics in mind, it is not surprising that more and more businesses turn to modern technology. The digital age offers both companies and consumers greater transparency, productivity, and comfort. At the same time, it presents companies despite new challenges.

16.6 EXAMPLE ON SUCCESSFUL USE OF CREATIVE MARKETING TECHNOLOGIES

In terms of the impact of technology, marketing is amongst the most prominent sectors. Vendors are always looking to engage an audience with creative ideas. We also endorse the use of creative marketing technologies at NewGen Apps. You can always visit our blogs covering big data uses in business, AI, and ML sales, retail marketing, etc. In this blog, we will share some of the boxes we use in marketing and sales technology.

Framestore VR Studio and Relevant developed a special teleportation experience of VR in collaboration with Marriott. Through creating a full immersion traveling experience for people the Framestore VR Studio tested technology boundaries. A concept was not limited by enhancing a vision and hearing. They took the classics a step further and built a vibrant advertisement. We updated our way of thinking with sensors and triggers within a stand-like phone framework. Oculus Rifts, heaters, and wind jets were their main toolkit.

They simulated a practical journey to Hawaiian and London with these stuffs. The tour included a choice of eight US cities with a period of 100 seconds each session. While it wasn't as nice as there, it could help you decide on your future journeys. It also helped Marriott position itself in the market as an odd and future-oriented brand.

The key point of sale was that it offered a balanced experience and went beyond a basic 360 picture. The idea was sufficiently successful for Marriott to begin his second "VRoom" initiative. We carried the virtual happiness of the hotel residents during this drive. People might take a VR headset from the room themselves and visit various locations around the world. In general, Marriott took a very creative move. Ideally, in the future we will see many more VR implementations.

16.7 CONCLUSION

Businesses are influenced by technological developments. Changes in technology pose obstacles, opportunities, and risks to organizations. Most businesses can use new technologies to improve their products and processes or even manufacture new ones to raise their sales and earnings. It can be an important tool for improving organizational processes and functions, or simply because it exists. This is how much the technology has to do with business. Company executives are always looking for technical changes and inventions. Such new technologies in the workplace-event-driven programming, knowledge increase, chatbots, and intelligent automation-these technologies are rapidly used to enhance business processes, efficiency for employees and employees 'experiences. In the coming years, it is hard to predict the full impact these new technologies will have on our lives, but it will certainly help to ensure an exciting and successful future.

The way companies operate in order to enable them to challenge large companies has been revolutionized by technology. The mere creation can be difficult to grasp with so many technological developments that take place so quickly. The mentioned above underlines some of the most important but far from detailed innovations. Once businesses are aware of technological innovation and adapt effective solutions to their particular needs and problems, growth, and success will become simpler than ever.

KEYWORDS

- 5G networks
- business environment
- cloud Computing
- competitive advantage
- gusto
- VR implementations

REFERENCES

Dedrick, J., Kraemer, K. L., & Xu, S., (2004). Information technology payoff in e-business environments: An international perspective on value creation of e-business in the financial services industry. *Journal of Management Information Systems, 21*(1), 17–54.

Gerstein, M., & Reisman, H., (1982). Creating competitive advantage with computer technology. *Journal of Business Strategy, 3*(1), 53–60.

Internet: http://www.informationr.net/ir/12-3/paper314.html (accessed on 21 October 2020).

Internet: https://www.gqrgm.com/top-10-business-trends-that-will-impact-growth-through-2022/ (accessed on 21 October 2020).

Internet: https://www.oksbdc.org/why-is-technology-important-in-business/ (accessed on 21 October 2020).

Internet: https://www.redalkemi.com/blog/post/impact-of-technology-on-business-environment (accessed on 21 October 2020).

Jasmine, M., Joshi, R. E., Chitra, K., & Ali, P. O., (2019). *Amity Global Business Review*, 32.

Krishnan, C., (2009). *Globalization and its Impact on Business Environment*. Indian MBA. com.

Krume, N., (2014). *The Role of Information Technology in the Business Sector International Journal of Science and Research (IJSR) ISSN (Online): 2319–7064, 3*(12).

Poolad, D., et al., (2010). Review of information technology effect on competitive advantage-strategic perspective. *International Journal of Engineering Science and Technology, 2*(11), 6248–6256.

Richa, G., Seema, S., Chitra, K., Gurinder, S., Chitra, B., & Priyanka, M., (2017). An empirical study to enquire the effectiveness of digital marketing in the challenging age with reference to Indian economy. *Pertanika Journal of Social Sciences and Humanities, 25*(4).

Seema, S., Richa, G., Priyanka, M., Chitra, K., Gurinder, S., & Chitra, B., (2018). *International Journal of Engineering and Technology, 7*(2), 52–57.

CHAPTER 17

Social Media Marketing and Purchase Behavior of Millennials: A Systematic Literature Review

JITENDER KUMAR,[1] SWETA DIXIT,[2] ALKA MAURYA,[3] and ASHISH GUPTA[4]

[1]Assistant Professor, Sharda University, Noida, Uttar Pradesh, India

[2]Associate Professor, Sharda University, Noida, Uttar Pradesh, India

[3]Professor, Amity University, Noida, Uttar Pradesh, India

[4]Assistant Professor, IIFT, New Delhi, India

ABSTRACT

The study presents a literature review on social media marketing and Millennials behavior. Twenty-six studies from reputed journals were analyzed searched to gain knowledge on SMM. This study helps in understanding the relationship between social media marketing and their impact on Millennials behavior accompanied by a comprehensive bibliography which will be helpful to both managers and researchers for studying existing research as well as for considering it for future research.

17.1 INTRODUCTION

The internet- and online-based social media have changed consumer consumption habits by providing consumers with new ways of looking for, assessing, choosing, and buying goods and services (Albors, Ramos, and Hervas, 2008). These modifications impacts how a firm works and influences marketing strategies by giving new challenges and difficult decisions (Thomas, 2007).

The rise of social media has changed the method of interaction between people and between the organizations and the people (Mangold and Faulds, 2009). Evolution of social media has changed the dynamics of the market. This rise of social media has threatened the competitors and has shifted the power to the consumers (Urban, 2005). The rise is evident from the increase in the active accounts among all social media platforms.

Facebook active monthly users worldwide as of 2nd quarter 2019 were 2.41 bn. Active users in social network sites has increased drastically, Facebook has 2410 Mn, YouTube has 2000 Mn, WhatsApp has 1600, Instagram has 1000 Mn active users. There are 270 million Facebook users in India alone which makes it leading country. Apart from India, there are other markets which have more than 100 million users like the USA, Indonesia, Brazil, Mexico, Philippines respectively (Statista, 2019). There are other SMM tools like Facebook, LinkedIn, Twitter, YouTube, WhatsApp, Microblogs like Twitter, Blogs, MySpace, Wikipedia, Virtual world, online forums, etc., are used to connect with customer in an innovative way (Mangold and Faulds, 2009; Perdue, 2010; Chan and Guillet, 2011). According to Kaplan and Haenlein (2010) defined social media as "a cluster of Internet-based applications that build on the ideological and technological foundations of Web 2.0, and allow the creation and exchange of user generated content." Social media cannot be comprehended without analyzing Web 2.0. Web 2.0 states a method wherein consumers utilize the WWW (World Wide Web), a venue where content is changed regularly by its operators in a collective and sharing way.

Organizations have presence through social links like page on Facebook and on other social networking sites to target their customers. Target customers increased with the help of social media ads. Company receives customer's online information as a form of their feedback through social networking sites emails (Nazir, Tayyab, Sajid, Rashid, and Javed, 2012).

Interaction among customers and the organizations develops brand loyalty (Jackson, 2011; Kaplan and Haenlein, 2010), that allows the promotion of goods, services also for the online networks of brand supporters (Kaplan and Haenlein, 2010). In addition, communication among customers gives firms new techniques for brand recall, awareness, and recognition (Gunelius, 2011).

According to Merlin David Stone, Neil David Woodcock, (2014) firms must build up its process, system capabilities, people so that they can communicate with their consumer, through omni-channel which customers want to use. Business intelligence should support to refine interactive marketing.

According to Castronovo and Huang (2012) methods like Public relations, promotions, Customer management product management in marketing should involve the use of social media because of the fact that consumer believes that content or the information posted on the social media platform is authentic and reliable as against the information provided by the organizations (Constantinides et al., 2010). Initially various organizations did include social media marketing as a part of their marketing strategy (Chan and Guillet, 2011) but later its seen that firms has accepted social media marketing for marketing proceedings like market research, customer relationship management (CRM), sales promotions, branding (eMarketer, 2013).

Social media permits to interact directly with brand or regarding brand with their family or friends. As internet has become part of the day-to-day activity for the majority. Social networking sites in recent years have gained importance years and have become a new way of communication. Its ability to connect individuals with each other has made it most effective and important business development tool (Tripathi, 2017). Social media promotes the social communication between consumers, which results in increasing trust and purchase intention (Hajli, 2104). Trust relies upon the apparent similitude between the one who provides the information and the user of that information. For instance, when somebody of 20 years suggests a specific drill, somebody of 65 Years will most likely pay attention to this exhortation; yet when somebody of 20 years prescribes a specific campground, the 65 years old individual will presumably not regard this counsel, as things that pull in more youthful people frequently repulse more seasoned individuals (Bronner, 2014). There have been many studies about online advertising; digital marketing but as far as the social media is concerned it has gained attention and notoriety among users, but remains new area for the academic researchers. There have been limited studies which are focused on the social media marketing for different customer age groups particularly for Millennials. There is a requirement of systematic study of literature to examine the intensity of these studies on Millennials.

Henceforth, our study extends and systemizes the study of social media and behavior of Millennials by reviewing the most conspicuous lines of research by organizing the conclusion of different social media marketing papers.

17.2 OBJECTIVES OF THE STUDY

- Review of existing studies on the social media marketing.
- To explore social media marketing and the change in the behavior of Millennials in the chosen papers.
- To find out research gaps in order to provide avenues for future research.

17.3 LITERATURE REVIEW

17.3.1 SOCIAL MEDIA MARKETING

Social media (SM) can create and develop linkages with customer through interactive firm and increasing the use of digital marketing by adopting various kinds of social media interactions which results in stronger customer relation, increase in customer engagement (Tiago, 2014). Customer relationship strategies can be developed by taking advantage of enhancement of customer engagement through social media (Tirunillai and Tellis, 2012; Leeflang, 2014). It does not only encourage communication between consumers to organization but also between customer to other customers (Mangold and Faulds, 2009). Firms have started adopting the social media platforms because they know that customers are closely associated, which results in increase in sales (Ho and Vogel, 2014). Consumers use social media like online communities to access user-generated content and to interact with other users (Hajli, 2014). Online brand communities play a significant role in influencing sales, attracting retaining the customers (Adjei et al., 2010). Relationship with customers can also be maintained with the help of SM (Porter and Donthu, 2008; Zhou et al., 2012; Brodie et al., 2013; Pereira et al., 2014).

Many studies has shown significant relationship between SMM and customer loyalty and study shows that SMM helps in developing loyalty among customers (Singh et al., 2008; Kaplan, 2012; Brodie et al., 2013; Kumar et al., 2013; Shih et al., 2014; Gamboa and Goncalves, 2014) and encourages customer engagement (Leeflang, 2014; Forbes, 2015).

The social media platforms, such as Facebook, Instagram, particularly Twitter, are highly preferred by Millennials (Sashittal et al., 2015). Keep the growing social media platforms, further study is expected whose impacts are able to impact the area in the coming years.

17.3.2 SOCIAL MEDIA MARKETING AND MILLENNIALS

Common span age range for Millennials changes is between 1981 and 199 (Jayson, 2010). Generation Y also called Millennials. Millennials are grown and brought up by socializing and the make online purchases, so usage of e-commerce for this generation will grow accompanied with its discretionary income (Smith, 2011). Technological products like Smartphone, computer are commonplace means for Millennials. In the USA, 97% of Millennials have computer and mobile phone are owned by 94% (Marketing Breakthroughs

Inc., 2008). Millennials uses digital media every day and have the capability to interact and purchase from a supplier anywhere in the globe. Because of their peer-to-peer communication and use of digital media, Millennials have become a motivating force in online shopping (Rehmani, 2011).

Users in the age group of 18–28 years had strong positive opinion for videos, blogs; brand channel ad formats (Cox, 2010). Digital media offers personalized relationship to consumers (Wind and Rangaswamy, 2001). Millennials replies to the personalized messages. Ads must focus to influence this age bracket, because this age group is special (Marketing Breakthroughs Inc., 2008). Customer loyalty towards the retailer can be enhanced through personalization (Srinivasan, Anderson, and Ponnavolu, 2002). To personalize relationship, online recommendations play vital role and which ranges from other customer's personal reviews to the recommender systems. Personalized information to the customer is provided by these recommender systems (Ansari, Essegaier, and Kohli, 2000). Many firms will be benefitted by giving personalized recommendations to online consumers. Consumers concentrate more on the recommendation sources as compared to the various category of websites on which these recommendations are posted (Senecal and Nantel, 2004).

There are approximately 378 million Millennials in India, which is higher in the world (The Indian Express, 2013). Because of this huge number of organizations are reconsidering their planning, strategies which are focused at Millennials. The famous brands at their parental generation are being refused by Millennials. Millennials are growing up in brand conscious and media saturated world as compared with their parents, there response to the advertisements is different (Neuborne and Kerwin, 1999).

17.4 METHODOLOGY

Existing literature on the social media was considered from 26 articles from the marketing journals from 2011 to 2019. For collecting references and literature required for our study, we started by researching the databases for the topic containing the term social media marketing and Millennials. We collected the research studies from Scopus and ABDC indexed journals having the social media marketing as an expression. Our analysis has been to confirm whether the study previously recognized were proper to the objective for our study. We held just those publications from English language journals that were not from conference proceedings.

Based on 26 articles we did systematic literature review and extracted summarized information to analyze their objectives, methodology, Finings limitations, and future research directions.

17.5 ANALYSIS AND DISCUSSION

The analysis segment of the study is categorized into two parts. The initial part manages with the patterns related to social media (literature classification), which provides the significant insights to the academia with respect to the recent research trends and patterns of the social media marketing. The later segment manages with social media marketing and the change in the behavior of Millennials used in the selected articles.

Studies revealed that social networks and microblogs are the most common social media adopted by firms and Millennials gives importance to the opinions of other fellow consumers and treat them credible than traditional tools of advertising (Smith, 2012). 5 studies adopted Facebook, 5 studies on Twitter, 5 studies on word of mouth, 2 studies on online advertising, 5 studies on customer engagement and 4 studies consumer behavior in general and purchase intentions in particular.

> **Time Period-Based Classification of Articles**
> Years
> 2011–3
> 2012–2
> 2013–3
> 2014–3
> 2015–4
> 2016–3
> 2017–2
> 2018–1
> 2019–5

These articles e*xplore social media marketing and the change in the behavior of Millennials.*

17.6 SOCIAL MEDIA MARKETING AND BEHAVIOR OF MILLENNIALS

Studies concentrated on millennial behavior, highlights two major areas of research (a) influence of social media on purchase intentions of Millennials (b) studies of the preferred social media marketing strategy by Millennials.

Purchase intention of consumer is a vital component while studying marketing but it consists of variables like social and attitude factors like consumer attitudes perceived value perceived risk, convenience, and the ease of use (Hidayat and Diwasasri, 2013). Ease of use to the customer is provided by digital marketing (Dehkordi and Javadian, 2012) changes consumer attitude (Cox, 2010; Ha, 2008). Social networking happens in the social media environment which has altered the methods to gather consumer's information and helps in buying decisions (Kaplan and Haenlein, 2010). Social media (websites, e-discussion, online chat, email, etc.) impacts the buying decisions of customers buying decisions. Study concentrated to examine the influence of social media on the purchase intention of the Millennials as the customer. Social media construct consists of seller created information and eWOM which has impact on customer's purchase intention (Rehmani, 2011). Due to social media a firm can anticipate the possibility of purchase intention. It depends on firm's discretion to choose between the social network (like Pinterest, Instagram, Facebook amongst others) and analyzing data of that particular network. Analysis of the social network's data enhances marketing endeavor of a firm by providing vital information about the user on the network and this also helps in understanding the social media strategy for that social network (Hill, Provost, and Volinsky's 2006).

Social media as an advertising strategy and it can be used as an effective tool to connect and communicate with their potential customers. Companies have started using social media at advanced level and have presence at all the platforms of social media (Neti, 2011). Effectiveness of social media relies on the quality of content, engrossment of the company, and the link with the advertising strategies (Pradiptarini, 2010). Social media is effective in influencing the behavior of Millennials (Smith, 2012). Millennials gives importance to the opinions of other fellow consumers and treat them credible than traditional tools of advertising. Word of mouth and Reviews of online consumers have the ability to reach hundreds and thousands of customers. As Millennials spend good amount of time online, companies can adopt right digital marketing strategies as an effective tool to communicate with this segment.

17.7 PREFERRED SOCIAL MEDIA MARKETING STRATEGY FOR MILLENNIALS

Social media platforms also includes online marketing communities, particularly, E-WOM, online ads, online communities which are efficient in enhancing brand loyalty and product purchase intention. To reach rising younger generation customers, social media marketing has become an important marketing tool (Balakrishnan, 2014). Coupons being the favorite mode of advertising for Millennials, superseded by email updates along with side panel ads, websites having graphics, bright colors, professional layout, and interactive site gains attention of this segment (Smith, 2012). Usage of social media has provoked the conviction that it is a significant tool which supports consumer engagement (Putter, 2017). Customer engagement helps in constructing emotional ties which can lead into WOM and can increase sales which is supported by range of merchandising options, shopping experiences and media options (Ha, 2008; Magneto, 2015; Forbes, 2015; Kannan 2017).

Organizations watches out new modes for reaching consumers and to shape consumer behaviors, including purchase intention and brand loyalty. Dynamic technology drives in growing activities in social networks like Facebook, YouTube, and Twitter, which have made courses through which consumers can create harmony, affinity, and start communications. Organizations embracing social network platform is probably going to be best in connecting, reaching, engaging, maintaining consumer base. Variables impacting brand perception and purchase intention comprise of the social component that leads consumer perspective and the viewpoint of others on social media posts. A developing vital fixation is on the utilization of user-generated content in light of specific brands and which impacts the view of other consumers. Brand value and purchasing decisions are associated more to the nature of peer interaction. Purchase intention may be established on perception that can be influenced by social media (Pütter, 2017).

17.8 IMPLICATIONS

With the help of study, we may recognize various implications of social media. One of the conclusions came to by the study implies how firms implement social media which are build up by social networks, like Facebook, Twitter (Chan and Guillet, 2011; Enli and Skorgerbo, 2013). There are many techniques for social media (Mangold and Faulds, 2009; Chan and Guillet, 2011). Another conclusion that can be drawn from the study is that organizations are joining millennials online and they are finding ways to establish a relationship with the millennials by pulling them to their websites.

Millennials are receptive to online coupons; an organization should make use include in online advertising broadly.

There have been various researches on online ad, digital marketing, but minimal research has been done preferred social media strategy by Millennials and which social media strategy impact their behavior. Finding of the study shows that there are social media strategies that can be effective in creating Millennials' attention and that influence the purchase intentions of millennials. Social media marketing encourage consumer engagement, helps in customer interaction with each other, it offers personalized relationship with the customers which can influence the consumer behavior and purchase intention of millennials, resulting into boost in the organization's sales.

This study will provide assistance to the managers to gain knowledge or customer insights about their perception, attitude, and behavior of Millennials, which has been explained through literature review process on numerous SM platforms, like Twitter, YouTube, Facebook, websites, and eWOM. This can assist managers to gain competitive edge. The managers are required to adopt promotional strategies for the SM platform of the company.

17.9 SCOPE FOR FUTURE RESEARCH

The study came across various limitations. Researchers have done major work on B2C but there is less focus on the B2B firms and the use of social media and limited research has been done as far as usage of social media in the B2B sector is concerned. Future studies could be analyzed with other generational groups like generation X. Future study of social media can be done on adoption and usage of SMM in the B2B segment. The study doesn't specify whether they are examining the social media adoption at the firm level or at functional level or at the individual level.

Most of the studies have concentrated either on examining the perspective of consumer on SMM to understand how they respond to these mediums, how organizations managing the relationships with their customer by extracting the extreme possible value from applications of such mediums. A study on exploring the organization's perspective when they use SMM, particularly regarding understanding the key barriers and obstacles to their usage should be taken under consideration.

KEYWORDS

- B2B segment
- marketing strategy
- millennials
- social media
- social media marketing
- worldwide web

REFERENCES

Alves, H., Fernandes, C., & Raposo, M., (2016). Social media marketing: A literature review and implications. *Psychology and Marketing, 33*(12), 1029–1038.

Baldus, B. J., Voorhees, C., & Calantone, R., (2015). Online brand community engagement: Scale development and validation. *Journal of Business Research, 68*(5), 978–985.

Behera, R. K., Gunasekaran, A., Gupta, S., Kamboj, S., & Bala, P. K., (2019). Personalized digital marketing recommender engine. *Journal of Retailing and Consumer Services.*

Boon, E., Pitt, L., & Salehi-Sangari, E., (2015). Managing information sharing in online communities and marketplaces. *Business Horizons, 58*(3), 347–353.

Brodie, R. J., Ilic, A., Juric, B., & Hollebeck, L., (2013). Consumer engagement in a virtual brand community: An exploratory analysis. *Journal of Business Research, 66*(1), 105–114.

Bunker, M. P., Rajendran, K. N., Corbin, S. B., & Pearce, C., (2013). Understanding 'likers' on Facebook: Differences between customer and non-customer situations. *International Journal of Business Information Systems, 12*(2), 163–176.

Chang, Y. T., Yu, H., & Lu, H. P., (2015). *Journal of Business Research, 68*(4), 777–782.

Diez-Martin, F., Blanco-Gonzalez, A., & Prado-Roman, C., (2019). Research Challenges in Digital Marketing: *Sustainability, 11*(10), 2839.

Facebook, (2015). *Facebook Reports First Quarter 2015 Results* [Online] http://investor.fb.com/release detail.cfm?ReleaseID=908022 (accessed on 21 October 2020).

Fierro, I., Cardona, A. D. A., & Gavilanez, J., (2017). Digital marketing: A new tool for international education. *Pensam. Gest., 43*, 241–260.

Ha, L., (2008). Online advertising research in advertising journals: A review. *Journal of Current Issues and Research in Advertising, 30,* 31–48.

Järvinen, J., Tollinen, A., Karjaluoto, H., & Jayawardhena, C., (2012). Digital and social media marketing usage in B2B industrial section. *Marketing Management Journal, 22*(2).

Kannan, P. K., (2017). Digital marketing: A framework, review and research agenda. *Int. J. Res. Mark., 34*(1), 22–45.

Kaplan, A. M., & Haenlein, M., (2010). Users of the world, unite: The challenges and opportunities of social media. *Business Horizons, 53*(1), 59–68.

Kaplan, A. M., (2012). If you love something, let it go mobile: Mobile marketing and mobile social media 4×4. *Business Horizons, 55*(2), 129–139.

Kumar, V., Bhaskaran, V., Mirchandani, R., & Shah, M., (2013). Practice prizewinner creating a measurable social media marketing strategy: Increasing the value and ROI of intangibles and tangibles for hokey pokey. *Marketing Science, 32*(2), 194–212.

Langan, R., Cowley, S., & Nguyen, C., (2019). The state of digital marketing in academia: An examination of marketing curriculum's response to digital disruption. *Journal of Marketing Education, 41*(1), 32–46.

Leeflang, P. S., Verhoef, P. C., Dahlström, P., & Freundt, T., (2014). Challenges and solutions for marketing in a digital era. *European Management Journal, 32*(1), 1–12.

Marketing Breakthroughs Inc., (2008). *Five Tips on Successfully Advertising to Gen-Y, Marketing breakthroughs Inc*. Available at: www.marketingbreakthroughs.com (accessed on 21 October 2020).

Paquette, H., (2013). *Social Media as a Marketing Tool: A Literature Review.*

Rehmani, M., & Khan, M. I., (2011). The impact of E-media on customer purchase intention. *International Journal of Advanced Computer Science and Applications, 2*(3).

Shashank, T. D. D., (2018). Effect of social media on online marketing: A study on young generation. *IOSR Journal of Business and Management*, 38–47.

Smith, K. T., (2011). Digital marketing strategies that Millennials find appealing, motivating, or just annoying. *Journal of Strategic Marketing, 19*(6), 489–499.

Stephen, A. T., (2016). The role of digital and social media marketing in consumer behavior. *Current Opinion in Psychology, 10*, 17–21.

Stone, M. D., & Woodcock, N. D., (2014). Interactive, direct, and digital marketing: A future that depends on better use of business intelligence. *Journal of Research in Interactive Marketing, 8*(1), 4–17.

Taken, S. K., (2012). Longitudinal study of digital marketing strategies targeting Millennials. *Journal of Consumer Marketing, 29*(2), 86–92.

Tiago, M. T. P. M. B., & Veríssimo, J. M. C., (2014). Digital marketing and social media: Why bother? *Business Horizons, 57*(6), 703–708.

Vinerean, S., Cetina, I., Dumitrescu, L., & Tichindelean, M., (2013). The effects of social media marketing on online consumer behavior. *International Journal of Business and Management, 8*(14), 66.

Wymbs, C., (2011). Digital marketing: The time for a new "academic major" has arrived. *Journal of Marketing Education, 33*(1), 93–106.

Yadav, M., & Rahman, Z., (2017). Social media marketing: Literature review and future research directions. *International Journal of Business Information Systems, 25*(2), 213–240.

CHAPTER 18

Eye-Tracking Analysis of Chosen Tourist Offers

MARIUSZ BARCZAK, PIOTR SZYMAŃSKI, and MARTIN ZSARNOCZKY

University of Economy/Emotin sp. z o.o./Kodolanyi Janos University

ABSTRACT

This study aims at analyzing the determinants that influence the selection of tourism offer. The objective was fulfilled by using eye-tracking analysis (with Gazepoint GP3 eye-tracker). A catalogs of six tour operators declaring the highest number of clients in 2006 have been selected for these studies. The study results indicate heat maps for the areas which were of interest for potential consumers. The findings shows that which elements of the catalogue attract the eye of potential consumers, i.e., how long they are watched, in what order, and which parts can be considered as focus areas. Studies have been conducted on subjectively chosen offers. Examined people were from one cultural environment and therefore applying the results on other nations is not possible. The results allow to introduce adjustments in the catalogues of tour operators. Even though eye-tracking surveys are increasingly used in marketing and advertising, they are very rarely connected with tourism industry.

18.1 INTRODUCTION

The constant development of information technologies, above all including mobile technologies, is one of the most important factors that contributed to changes on tourism services market. Traditional catalogs issued by tourism companies are increasingly being replaced by websites, often available in mobile versions. Nevertheless, there still is a group of buyers for whom the printed catalog is the main source of information about tourist destination.

The survey conducted by Center of Public Research in 2009 on the sample of 1000 respondents reveals that the decision of purchase is most often made in the travel agency during the conversation with the seller (63%). Therefore, the article aims at analyzing the selected tourism offers by the use of eye-tracking analysis.

18.2 OVERVIEW OF LITERATURE

Focus on services, and, as a consequence, immateriality of tourism is a reason why the industry is in majority based on images, such as photographic images which attract attention of recipients and are crucial for choosing offers (Wang and Sparks, 2014, pp. 588–602). Publications concerning eye-tracking in 2019 include, among others, websites, and travel blogs (Muñoz-Leivaa, Hernández-Méndez, and Gómez-Carmonac, 2019), analysis of ratings on reservation platforms (Coba, Zanker, and Rook, 2019, pp. 40–51), studies on new trends and innovations in marketing of tourism services (Bigné and Decrop, 2019, pp. 131–154).

In recent years in Poland the analysis of determinants which have an impact on choosing the tourism offer was conducted, among others, by Szlachciuk and Ozimek. They focused on the quality of tourism services provided by tourism agencies (Szalchciuk and Ozimek, 2016, pp. 216–227). The aim of the quoted article was to determine the impact of selected factors on the choice and assessment of the quality of travel agencies, as well as on perception of the quality of tourism services. The researches were conducted in 2012 among 1000 people aged 15 to 70 from Poland who used the services of tourism agencies. However, they were not referring to visual aspects of the offer. Michalska-Dudek and Rapacz (2015, pp. 141–152), on the other hand, addressed the issue of measuring the loyalty of travel agencies customers. They emphasized the fact that a relation between a consumer and a tourism enterprise is very complex and diverse. Understanding the consumer of tourism services requires the use of various methods and marketing tools that are not only in the area of interest of science, but above all business practice. Researchers were also interested in behavior patterns of tourism website users (Gontar, 2016, pp. 313–326), attempting an answer on questions concerning factors that are taken into account during decision-making. On the basis of the studies conducted with the use of interview questionnaire it was agreed that the most important factors include: the price (21.5% of respondents), description of the facility (14.5%) and the room-decorations and cleanliness (13.3%).

Only 8.3% of respondents emphasized the importance of the photographs of the facility and its surroundings (Gontar, 2016, p. 319).

The studies with the use of eye-tracking analysis were also conducted, among others, by Jerzyk (2017, pp. 122–131), Garczarek-Bąk (2016, pp. 54–71), Wąsikowska (2015, pp. 177–192) who focused on visual attention of the consumers, analysis of buying behavior determinants and marketing analysis. The use of eye-tracking analysis in tourism was, among others, limited to identification of the landscape elements (Młynarczyk and Potocka, 2010, 2011) that attract the eye. Thus, the authors evaluated 18 photographs (8 photographs of various types of landscape). Pilot studies with the use of mobile eye-tracker were also conducted in subsequent years.

18.3 SITUATION OF TOURISM COMPANIES IN POLAND

According to the data of Ministry of Sports and Tourism (MSiT), 57% of the residents of Poland aged 15 years or over participated in at least one tourist travel. The growth was observed in long-term travels (from 15.7 million in 2015 to 17 million in 2016), as well as in short-time (from 24.2 million in 2015 to 26.5 million in 2016).

Most frequently visited foreign destinations in 2016 included Germany (2.3 million), Italy (0.9 million), Great Britain (0.9 million), Croatia (0.91), Czech Republic (0.7) and Greece (0.62 million). MSiT estimated that the average stay lasted approximately 9 nights, with as many as 66% of tourists organizing their travel individually. Travel agencies were chosen only by 25.1% (with 26.8% in 2014 and 25.5% in 2015). Choosing the means of transport, the majority of clients favored a plane-86.4%, their own transport-6.9% and a coach-6.4% (Table 18.1).

TABLE 18.1 Goals of Domestic Tourist Trips of Polish Inhabitants in 2016

Trips/Goals	Leisure	Visit	Business	Health	Religion	Others
Short-term	27.3%	61.4%	4.1%	1.0%	2.9%	3.3%
Long-term	58.0%	31.6%	2.2%	6.0%	0.7%	1.5%

Source: Turystyka w Polsce (2016). https://msit.gov.pl/pl/turystyka/badania-rynku-turystycz/7401,Turystyka-w-Polsce-2016.html.

The most popular destinations in 2016 included Greece (33.0%), Spain (18.2%), Bulgaria (9.0%), Turkey (6.9%), and Italy (6.2%). The market, however, proved to be very sensitive to changes in environment, including above all political or terrorist issues. As the result of the abovementioned factors, the Egyptian market decreased by 66.8%, Turkish by 58.3%, and French by 39.9%. The most popular Greece destinations were: Heraklion (7.8% of the market in 2016), Burgas (5.8%), Corfu (5.4%), Zakynthos (5.2%), Rhodes (5.0%) (Table 18.2).

TABLE 18.2 Declared Number of Clients of Tour Operators in Poland

Tourism Company	Number of Clients		
	2016	2015	2016/2015
Itaka	637000	607000	+4.9%
Rainbow Tours	330300	300200	+10.0%
TUI Poland	318260	312000	+2.0%
Wezyr Holidays	135370	189406	−28.5%
Grecos Holiday	168500	137000	+23.0%
Neckerman	119884	142594	−15.9%

Source: Touroperatorzy: Turystyka Wyjazdowa. Ranking (2017), Wiadomości Turystyczne 8/2017, p. 15.

18.4 RESEARCH

Based on the analysis of the situation of tourism companies in Poland, catalogs of six tour operators declaring the highest number of clients in 2006 have been selected for the test (Table 18.2), i.e., Itaka, Rainbow Tours, TUI Poland, Wezyr Holidays, Grecos Holiday, Neckerman. The other criteria taken into account were the most popular countries and destinations. Therefore, the chosen destination was Greece, Korfu. The choice was made with the use of purposive sampling, where the researcher indicates the test entities on the basis of one's own knowledge (Szreder, 2010, pp. 56, 57). Due to the focus on the variety of phenomena, not on the frequency of their occurrence, the chosen method was the qualitative research. It is used to identify the opinion, preferences, and attitude towards market phenomena and processes (Ruszkiewicz, 2002, p. 29). Qualitative researches provide information about the ways in which the respondents perceive and evaluate certain phenomena and processes, and explain the mechanism of how certain opinions are created (Kędzior, 2005, p. 83). The research sought to address the following questions: 1) what elements attract the attention of potential clients in the first place? 2) Is it possible to define the scheme of offer browsing on the basis of the research results?

18.5 CONDUCT OF THE STUDY

The research was conducted on a test bench equipped with the monitor (17 inches), laptop (Lenovo Yoga) and stationary eye-tracker (Gazepoint GP3). The researcher sat on the right side of the respondent, controlling the test (Figure 18.1).

FIGURE 18.1 Test bench settings.
Source: Own study.

The respondent was situated according to the scheme presented in Figure 18.2. The essential part of the research started after device calibration. The researcher read the instruction of the research conduct.

The study was conducted among 30 students aged 19 to 24, including 20 women and 10 men. The majority of respondents indicated that they travel three times a year (17 people), then 2 times a year (7 people) and once a year (6 people). The most common aims of the trip were leisure (20 people) and sightseeing (10 people). None of the respondents indicated health or religious reasons. The last question regarded the trip organizer. In this case, the majority of respondents indicated that they do it individually (19 people), whereas 11 people indicated travel agencies (traditionally or via internet).

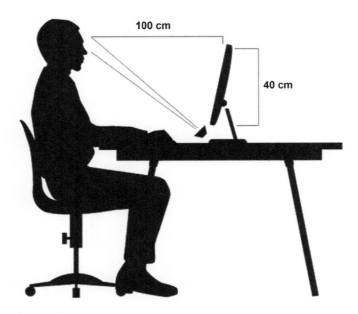

FIGURE 18.2 Test bench settings.
Source: Own study.

Analyzed were electronic copies of offers of such tourism companies as Itaka, Rainbow Tours, TUI Poland, Wezyr Holidays, Grecos Holiday, Neckerman (Figure 18.3).

The respondent was looking at the offers for the periods of 10 s. Therefore, the total exposure time of all offers was 60 s. The offers were displayed randomly. Short time of display allowed to compare the results from different observers and to indicate elements that attracted the eye of the observer who saw it for the first time.

Collected material was examined with the use of frequency analysis. The results were illustrated with heat maps with 2 s. time interval. Thus, it was possible to present the results taking into account the time of exposure. The results were also presented for the whole 10 s. time of offer exposition.

Eye-Tracking Analysis of Chosen Tourist Offers 299

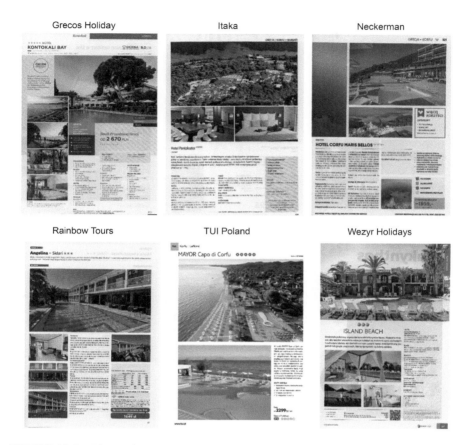

FIGURE 18.3 Electronic copies of offers.
Source: Own study.

18.6 RESULTS

Figure 18.4 presents the results of analysis as heat maps which depict the areas of eye-focus of the observer. It is an overview of the frequency of glances in the whole 10 s. interval. Warm colors (dark red) show the maximal eye-focus. Each illustration has been characterized below:

FIGURE 18.4 Areas of eye focus (the entire period).
Source: Own study.

1. **Grecos Holiday:** Most frequently, the respondents looked at graphic elements and the price in the middle part of the offer. Texts in the lower part attracted relatively low attention. Similarly, the eye was not caught by the title of the offer in the upper part.
2. **Itaka:** The gaze of the respondents was concentrated in the center. Especially interesting were two illustrations in the middle. The respondents read the texts partially. A relatively significant number of people found the price in the lower right corner. The title of the offer in the center attracted the eye of the respondents relatively often.

3. **Neckerman:** Graphic elements in the upper and middle part attracted the eye. The respondents scanned also the text below. In the lower right corner the price attracted the attention. The title of the offer placed in the middle, resulted in a relatively high gaze concentration.
4. **Rainbow Tours:** Without doubt, graphic elements in the right upper corner of the offer aroused the most interest. Illustrations in the lower part and on the right were definitely less explored, with the exception of the price in the lower right corner. The respondents did not pay much attention to the tile of the offer at the top.
5. **TUI Poland:** The offer consisted of three noticeably separated parts. Graphic elements and the price in the lower right corner attracted much attention. Few of the respondents paid attention to the title of the offer at the top.
6. **Wezyr Holidays:** The respondents focused their eyes in the center of the offer noticing the title. Texts on the right side also attracted much attention.

Figures 18.5 and 18.6 illustrate heat maps of each offer for two-second intervals. Therefore, each illustration depicts heat maps with the data recorded every 2 seconds. Because of this focus visualization method it was possible to present the areas of gaze concentration in dynamic perspective.

The particular offers can be characterized in the following way:

1. **Grecos Holiday:** The respondents started the observation in the upper left corner (name of the hotel, text, swimming pool, beach). Then, their sight was directed to the right side (swimming pool infrastructure) or to the lower left side (beach, swimming pool, rooms). In the next period, the majority of eye focus is concentrated in the center of the offer (price, texts). Concentrations are dispersed.
2. **Itaka:** The observers started with focusing the eye on the elements in the upper right corner (red roofs of the houses). Then, they moved their sight to the center (table, bed with flowers) and finished their exploration in the lower part (texts of the offer and the price). Concentrations are dispersed.
3. **Neckerman:** At first, the respondents focused their eye on the upper right quarter of the offer (swimming pool, the view). Then, they moved their sight to the center (the view, more benefits) to finish watching in the lower part (the view, price). Concentrations are dispersed.

302　　　　　　　　　*Integrating New Technologies in International Business*

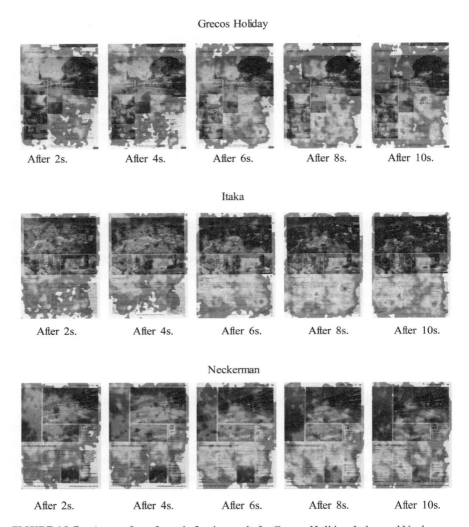

FIGURE 18.5　Areas of eye focus in 2 s. intervals for Grecos Holiday, Itaka, and Neckerman offers.
Source: Own study.

For the subsequent offers the characteristics are:

1. **Rainbow Tours:** The respondents started in the upper part (swimming pool, buildings near the swimming pool), moved to the center of the left side (rooms, texts) and finished in the lower part (offer, price). Concentrations are dispersed.

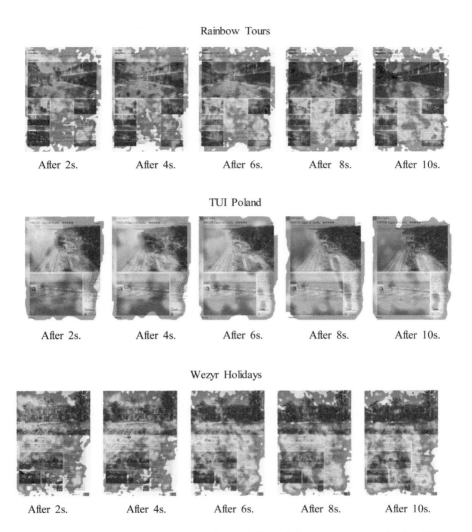

FIGURE 18.6 Eye concentration in 2 s. intervals for Rainbow Tours, TUI Poland and Wezyr Holidays offers.

Source: Own study.

2. **TUI Poland:** The respondents acted in a more systematic way. They started with the biggest graphic element at the top (room, buildings). After that, they watched the left part (swimming pool) and finished on the elements in the lower right side (descriptions, price).

3. **Wezyr Holidays:** Browsing starts with the lower upper corner (swimming pool and surroundings, text). Following it is the middle part (text, swimming pool). Browsing ends in the lower middle and right side of the offer (swimming pool, descriptions).

The analysis of research problems allows saying that the respondents paid attention to graphic elements at first. What is more, the majority of the respondents viewed the offers in the shape of letter F, i.e., from the upper to the lower part, and from left to right. It is a very typical behavior, resulting from reading habits in western world.

18.7 CONCLUSIONS

The conducted analyses proved that in the majority of cases the sight of the respondents is drawn by graphic elements and a significant number of various elements results in dispersion of the eye focus. The respondents viewed the offer according to the well-known F shape. It means that the exploration starts in the upper left corner, then the respondents focus their sight on the elements at the top and finish on the elements in the lower left side. The prices proved to be the element that always attracts the attention. The lack of price resulted in the increased interest in other elements.

From among the analyzed offers, the TUI Poland proved to be observed in the most schematic way. It is a noticeably distinguished offer amid the others in terms of the sequence and repeatability of eye focus on certain parts. In this case, at first the focus was on pictures, then on the text and finally on prices.

The conducted analyses may be the source for preparing an optimal and effective marketing strategy with the use of catalogs. The results will allow introducing possible adjustments in the design of new catalogs. It is the visual message that often creates the tourist need encouraging to future consumption. A descriptive part is only addition and clarification of this message.

Even though eye-tracking researches have been increasingly used in marketing, they are rarely connected with the tourism industry. Expansion of the scope of this type of research in connection with the analyses of websites presenting such offers, also enriched with quantitative research, should be continued in the future.

KEYWORDS

- eye tracking
- graphic elements
- oculography analysis
- tourist supply
- Wezyr holidays

REFERENCES

Bigné, E., & Decrop, A., (2019). Paradoxes of postmodern tourists and innovation in tourism marketing. In: Fayos-Solà, E., & Cooper, C., (eds.), *The Future of Tourism* (pp. 131–154). Springer.

Bogucki, Wydawnictwo, N., Poznań, Młynarczyk, Z., & Potocka, I., (2011). Możliwości wykorzystania eye-trackingu w badaniach krajobrazu turystycznego, [in:] Przestrzeń turystyczna–czynniki, różnorodność, zmiany, WGiSR, Uniwersytet Warszawski, Warszawa.

Coba, L., Zanker, M., & Rook, L., (2019). Decision-making based on bimodal rating summary statistics: An eye-tracking study of hotels. In: Pesonen, J., & Neidhardt, J. (eds.), *Information and Communication Technologies in Tourism*.

Efekt ROPO w-segmencie Travel, (2009). TNS OBOP, http://docplayer.pl/storage/26/7437596/1561191745/ISZ4dzBh7100fyxKaXduHw/7437596.pdf (accessed on 21 October 2020).

Garczarek-Bąk, U., (2016). Użyteczność badań eye trackignowych w pomiarze utajonych determinant zachowań zakupowych nabywców, Ekonometria / Uniwersytet Ekonomiczny we Wrocławiu, Wrocław..

Gontar, B., (2016). Wzorce zachowań użytkowników stron turystycznych, Roczniki Kolegium Analiz Ekonomicznych / Szkoła Główna Handlowa, Warszawa.

Jerzyk, E., (2017). Zastosowanie okulografii w badaniach uwagi wzrokowej konsumentów, Handel Wewnętrzny, Warszawa.

Kędzior, Z., (2005). Badania rynku: Metody, zastosowania, PWE, Warszawa.

Michalska-Dudek, I., & Rapacz, A., (2015). Pomiar lojalności usług biur podróży w Polsce, *Studia Oeconomica Posnaniensia*, 5(4), 242–264.

Młynarczyk, Z., & Potocka, I., (2010). Próba obiektywizacji oceny krajobrazu turystycznego przy wykorzystaniu zjawiska sakad. In: Młynarczyk, Z., & Zajadacz, A., (eds.), Uwarunkowania i plany rozwoju turystyki, Turystyka i Rekreacja – Studia i Prace, 7.

Muñoz-Leivaa, F., Hernández-Méndezb, J., & Gómez-Carmonac, D., (2019). Measuring advertising effectiveness in travel 2.0 websites through eye-tracking technology. *Physiology and Behavior, 200*, p. 83–95.

Polska, T., (2015/16). z perspektywy PZOT (2016).

Poznań, & Michalska-Dudek, (2015). Badanie lojalności klientów biur podróży, Zeszyty Naukowe Uniwersytetu Szczecińskiego. Problemy Zarządzania, Finansów i Marketingu, no. 41, T.2, Szczecin.

Ruszkiewicz, M., (2002). Metody ilościowe w badaniach marketingowych, PWN, Warszawa.
Szlachiuk, J., & Ozimek, I., (2016). Jakość usług turystycznych świadczonych przez biura podróży w Polsce, Handel Wewnętrzny, (5).
Szreder, M., (2010). Metody i techniki sondażowych badań opinii, PWE, Warszawa 2010.
Touroperatorzy: Turystyka wyjazdowa. Ranking 2017, Wiadomości Turystyczne 8/2017.
Turystyka w Polsce, (2016). https://msit.gov.pl/pl/turystyka/badania-rynku-turystycz/7401, Turystyka-w-Polsce-2016.html (accessed on 21 October 2020).
Wang, Y., & Sparks, B., (2014). An eye-tracking study of tourism photo stimuli: Image characteristics and ethnicity. *Journal of Travel Research*, 1–15.
Wąsikowska, B., (2015). Eye tracking w badaniach marketingowych, Zeszyty Naukowe Uniwersytetu Szczecińskiego. Studia Informatica, Szczecin.

Index

A

Aadhaar enabled payment system (AEPS), 177, 178
Acquisitions, 43, 46
Acquisitiveness, 98
Adjusted goodness of fit (AGFI), 22, 103, 113, 115
Administrative functions, 139
Affective dimension, 123, 128
Agile competition, 151
Agreeableness, 61, 62, 66–68
Agricultural era, 237
Agriculture extension, 191
Agritech startup, 199
Alpha reliability, 23
Amazon, 51, 163, 164, 178, 183
  pay, 178, 183
Ambiguity, 38
American marketing association (AMA), 211
Analyze, model, process, execute, evaluate (AMPEE), 153, 167–169
Android pay, 182
Anthropology, 41
App store, 163
Arbitrating authority, 137
Artifacts, 36, 37
Artificial
  environment, 255
  intelligence (AI), 71–93, 187, 191, 192, 203, 233, 234, 236, 275, 277
    recent developments, 81
    technologies, 74–78, 80, 82–93
  intelligent device, 93
Augmented reality (AR), 253–268
Authentic communication, 254
Automation, 71, 77, 79, 80, 90, 156, 158, 164, 233, 235, 237, 248, 250, 271, 278
Average variance explained (AVE), 23, 24, 105, 111–113

B

Backward extension, 201, 203
Balance sheet insolvency, 131
Balanced score card, 153
Bank pre-paid cards, 179
Banking
  cards, 177, 178
  channels, 128
  touchpoints, 127
Bankruptcy, 131, 132, 135, 136, 142
  code, 135, 138, 143
Barometer, 10
Behavioral dimension, 125, 128
Beta coefficient, 25, 26, 62–65, 67
Bharat interface for money (BHIM), 181, 182, 184, 185
  cashback scheme, 181
  promotional strategies, 181
Big
  data, 14, 77, 233, 277
  five
    model, 66, 68
    personality, 58
BigDog, 75
Bi-level verification, 178
Biometric, 83
  authentication, 178
Biotechnologies, 191
Black
  box technology, 199
  money, 176
Block, 225, 226, 230
Blockchain, 225–231
  transaction, 229
Bottom-up approach, 73
Brain drain, 86
Branch-banking, 119
Brand
  awareness, 97, 99
  communication, 267
  conscious entities, 101
  consciousness, 99, 110, 113

engagement, 254, 255, 259, 266, 268
image-dress code, 37
British
   textile employees, 249
   virgin islands, 141
Bureaucracy, 38
Business
   communications, 149
   domains, 46, 48
   ecosystems, 150, 154, 164, 165
   enterprise, 1, 9, 10, 151, 155, 157
   environment, 32, 33, 42, 43, 45, 48, 151, 154, 157, 240, 247, 269, 271, 279
   foundation, 212
   goals, 152
   growth, 149, 150, 158, 274
   income, 7
   innovation, 149–154, 159, 165–170, 244, 265
      context, 150
   leadership, 151, 154, 155
   literatures, 151
   management, 42, 154
   model (BM), 1–8, 10, 11, 72, 149–154, 158, 159, 163–170, 187, 234, 238
      demographic characteristics, 6
      wealth and welfare, 6
   organization, 234, 235, 237, 241, 247, 251
   performance evaluation (BPE), 153
   practice, 32, 157, 294
   process
      framework, 153
      management (BPM), 152, 153
   researchers, 152
   sector, 205, 214, 269–271
   strategy (BS), 1, 149, 152–156, 158, 168, 233, 237, 238, 244–246, 251
   transformation, 149–159, 170
   units, 169
   value chain (BVC), 153

C

Capital-structured theory, 6
Carbon dioxide (CO_2), 196, 198
Cashless
   economy, 174, 175, 177, 179, 185
   payments, 174
   transaction, 176, 185

Centre of main interest, 136
Chinese communities, 96
Cisco, 15, 154
Clerical employees, 248
Client servicing individual, 83
Cloud computing, 72, 242, 276, 279
Co-evolutionary analysis, 4
Cognitive dimension, 122, 123, 128
Cold chain infrastructure, 190
Collectivism, 38
Collision, 234
Commercial
   footprints, 46
   interests, 47
Communal tools, 241
Communication
   era, 237
   styles, 34, 42, 43
Comparative fit index (CFI), 22, 103, 113, 115
Competitive advantage, 48, 149, 150, 152, 159, 238, 244, 246, 275, 279
Composite reliability (CR), 23, 111–113
Conceptual model, 101, 119, 121
Confirmatory factor analysis (CFA), 13, 19, 21–23, 27
   model fit indices, 22
Conflict resolution, 43
Connoisseurship, 95
Conscientiousness, 62, 63, 66, 67
Consumer
   behavior, 256
   generated media, 107
Contractual principles, 134
Controlled atmosphere (CA), 196–198, 203
Conventional
   advertising, 261, 266
   mass-market players, 4
Cook's distance method, 20
Cordial negotiations, 47
Corporate
   communication, 43
   culture, 33, 34, 43
   debtor, 138
   social responsibility (CSR), 1, 6, 7, 9–11, 43
Cost
   and privacy, 13, 22
   effective storage, 72

optimization, 158
reduction, 159, 165, 230, 254, 260
Craftsmanship, 95
Credit card, 183
Cronbach's alpha, 23, 58, 104, 110
Cross
  border
    insolvency, 132, 135, 136
    framework, 135
    regulations, 142, 146
      investments, 132, 150
      nature, 132
      pauper, 132
      pauperdom, 135
      transactions, 132
  cultural
    advantages, 51
    amalgamations, 51
    association, 51
    business environments, 43
    communication, 37, 43, 44, 47
    competence, 32
    context, 40
    diversities, 42, 47, 50, 52
    integrations, 42
    knowledge, 50
    management, 34, 35, 41, 42, 49, 52
    negotiations, 44
    organizations, 45
  cross-cultural communication, 47
  expatriate manpower and local human capital management, 46
  global business environments, 45
  intercontinental strategy, 46
  intercultural mentorship, 46
  multicultural conflict resolution, 47
    psychology, 41
    relationships, 33
    roadblocks, 43
    sensitivity, 50
    teams, 32, 33, 47, 48
    understanding, 40, 50, 51
    workforce, 48
  culture, 31–33, 40–42, 47, 49–52
    advantages, 50
    amalgamations, 42
    business environments, 32
    management, 31, 32, 42, 47

fringe
  indebtedness, 132
  nature, 132
functional teams, 157
outskirt
  bankruptcies, 132
  indebtedness, 137
sectional data, 19
Cryptographic trust, 230
Cultural
  adaptation, 48
  awareness, 35
  barriers, 32, 47
  business influences, 47
  change, 37
  distance, 52
  diversity, 49, 50
  Edgar Henry Schein model, 36
  integration, 50
  levels, 35
    institutional level, 35
    psychic level, 35
  perspective differences, 35
  sentiments, 210
Customer
  bank relationship, 26
  convenience, 254, 258, 265, 267
  engagement, 257, 265, 267, 288
  experience, 14, 77, 119–128, 157, 158, 244, 255, 266, 267
  dimensions, 120
  loyalty, 126
  relationship management (CRM), 1, 13–15, 19–22, 24–27, 237, 283
  satisfaction, 126, 165, 222, 265
Cutthroat competition, 266
Cybercrimes, 82, 83
Cybernetic zone, 107
Cybersecurity, 82, 86

D

Data
  analysis, 4, 5, 103, 275
  collection, 4, 18, 207
    instruments, 4, 5, 19
  privacy, 26
  screening, 19, 20
Debit card, 178

Debtor company, 144
Decision-making, 42, 50, 120, 261, 272, 294
Deep
　blue, 74
　learning, 93
Degree of job automatability, 81
Deloitte's analysis, 245, 246
Demographics, 50, 56, 68, 103
Demonetization, 173, 174, 176, 179, 181, 185
Dependent variables, 3, 9
Digital
　and communication network industry, 166
　archive online, 276
　business
　　model, 165, 170
　　transformation, 154
　capabilities, 150, 151, 154, 157, 163, 166, 169
　channels, 260
　culture, 157
　dimensions, 153, 154
　economy, 154, 157, 179, 184
　elements, 267
　era, 237
　growth, 179, 182
　India, 173, 185
　infrastructure, 93
　innovation, 169, 190
　interface, 274
　ledge technology, 231
　marketing, 253, 287
　matrix, 154
　mechanism, 175
　media, 159, 268, 285
　mobile wallet, 178
　mode, 125, 174
　payment, 173–180, 183–185
　　applications, 175, 183
　　drawback, 184
　　mode, 177
　　operators, 174
　　strengths, 183
　　system, 175–178, 183
　platforms, 150, 154, 157, 166, 259, 265, 266
　proliferation, 154
　resources, 154
　samples, 258

smart cards, 177
supply chain solutions, 202
technologies, 153, 157, 158, 165
transactions, 174, 182, 184, 225
transformation, 153, 154, 157–159, 164–166, 168–170, 274
wallet, 174, 178, 180, 182–184
Digitalization, 149, 158, 150, 151, 157–159, 165, 168–170, 173, 203, 255, 262, 263, 267, 268
Digitization, 85, 90, 150, 157–159, 163, 165, 175, 193, 202
Dimension caters, 123
Disruptive technology, 71, 72, 93
Distributed database, 225, 231
Domestic
　inland transport, 144
　marketing, 212, 213
　markets, 210
Dosage cycle, 77
Draft part Z, 137, 143, 145
Dynamic, 4, 96, 100, 152, 166, 203, 253, 254, 282
　changing environment, 92
　managers, 42
　process network, 272

E

E-commerce, 113, 173, 179, 284
　shopping, 173
Economic
　health, 132
　uncertainty, 2
Economical advertising, 263
Ecosystem, 48, 164, 166, 191, 222
Edgar Henry Schein's organizational model, 36
E-discussion, 287
Education sectors, 92
Effective communication, 267
Eigen standards, 105
E-learning, 14
Electric stackers, 199
Electronic
　banking, 14
　clearing system, 178
　payment system, 182
　transactions, 173

Elucidation, 132, 137
E-mail questionnaire, 58
Emotional
  intelligence, 57
  stability, 56, 57, 64–68
Empirical research, 97
Employee
  conscientiousness level, 62, 63
  demographics, 68
  openness, 65, 66
  personality, 55, 68
    characteristics, 68
Employment, 78, 103, 108, 233, 244, 247–251, 275
  technology interaction, 250, 251
Encryption technique, 226
Enterprise
  dimensions, 167
  resource planning, 237
Entrepreneur, 1, 3, 5, 7–11, 85, 191
Entrepreneurship, 149, 151, 152, 167, 168, 192
Equivalent qualifications, 59
Ethnicities, 44
Ethnocentrism, 33, 43, 52
Etiquettes, 43, 52
E-touch points, 125
E-trust, 125
European insolvency regulations (EIR), 142, 146
Experiential marketing, 266
External organizational networks, 37
Extraordinariness, 102
Extraversion, 57, 59–61, 66–68
Eye tracking, 305
  analysis, 293, 295

F

Facebook, 96, 98, 164, 180, 263, 264, 282, 284, 286–289
Factor analysis, 37, 104
Fanatical aggressive behavior, 45
Fanciful era, 244
Financial
  market fluctuations, 42
  misappropriation, 201, 203
  risks, 136
  ruination definition, 131
Fingerprint authentication, 178

First order
  business dimensions, 153
  CFA, 21
Flipkart, 182
Flying shuttle, 249
Food
  chain, 227
  industry, 195
  monitoring, 195
  security, 231
Foreign
  domains, 48
  main proceeding, 138
  non-main proceeding, 138
Forex, 43
Forging manufacturing, 45
Fourth industrial revolution, 274, 275
Fraudulent transactions, 229
Freecharge, 178
Fresh and healthy enterprise limited, 196, 197

G

Gamification, 257, 267
Gazepoint, 293, 297
Gender profile, 58
General linear modeling (GLM), 55, 58, 68
Global
  advertising, 205, 207, 211, 215, 218, 264
  barriers, 32
  brains, 50
  business, 33, 45–49, 51, 52, 71, 164, 222, 263
  communities, 50
  competence, 46
  competitive edges, 33
  compliance norms, 33
  cultural awareness, 35
  culture, 50, 51
  digital statshot report, 263
  domains, 221
  economy, 49, 52, 134, 145
  exchange, 132, 215
  expansion, 48, 50
  firms, 43, 208, 219
  food security, 231
  footprints, 42
  framework, 136
  interdependent economy, 134

labor market, 242
landscape, 210
managers, 42, 44
market, 51, 205, 206, 219, 221, 222, 254, 269, 270
   digitalization, 263
mobility, 210
multicultural business communities, 34
networking, 50
organizations, 41, 42, 48
outlook, 47
outreach, 33
partnership, 51
promoting, 205, 207
selling, 205
showcasing, 206–208, 211, 216
standards, 135, 145, 192
trading, 206
work practices, 32
working teams, 41, 44, 49
workstations, 50
Globalization, 32, 42, 47, 49–51, 132, 150, 205–208, 210, 212, 216, 218, 221, 222, 272
Golden state foods, 225
Good-fit model, 22
Goodness of fit (GFI), 22, 103, 113, 115
Goods and services, 132
Google, 18, 51, 76, 82–84, 164, 182, 236, 241, 259
   assistant, 83
   glass, 18
   pay, 182
Graphic, 6, 254, 288
   elements, 300, 301, 304, 305
Grecos holiday, 296, 298, 300–302
Greenhouse, 191
Gusto, 277, 279

H

Happiness index, 43
Harman's one-factor test, 21
Harmonization, 143
Harvard business review, 156, 164
Healthcare
   facilities, 76
   sector, 76, 77, 82, 92
Hedonic elements, 124

Hierarchical
   conceptualization, 150
   dimensions, 150, 151, 167
   linear regression, 60, 68
High
   court, 141–145
   power index, 38
   uncertainty, 38
Hindustan Unilever limited (HUL), 51
Hofstede's
   cultural dimension, 37
      model 1, 40
      theory, 37
   six-dimensional models, 40
Holy Quran, 10
Homogeneity, 210
Horizontal
   coherence, 4
   value chains, 163
Horticulture, 187, 188, 190
Household showcasing, 214, 215, 221
Human resource management, 237
Humanoid, 91
   robots, 75
Hybrids, 192

I

Ideation, 46, 155
Image management, 43, 48
Immediate payment service, 178
Income insolvency, 131
Incubations, 46
Independent variables, 3, 8, 9
In-depth interviews, 1, 5
Indian
   draft, 138
   economy, 135, 145, 176
   insolvency laws, 133
   market, 90, 145, 183, 193, 208, 212
Individualism vs. collectivism (IDV), 37, 38, 52
Indulgence vs. restraint (IND), 37, 38
Industrial era, 237
Industrialization, 221
Influx, 41, 42
Infographics, 83

Index

Information
　and communication technology (ICT), 202, 203, 237
　architecture, 128, 231
　technology (IT), 14, 17, 27, 90, 155, 156, 164, 166, 169, 191, 199, 233, 237, 241, 245, 250, 255, 269–271, 273, 275
Informativeness, 98
Innovation
　capabilities, 151, 167, 169
　management, 155
　　initiatives, 155
　　standards, 151
　　sub-construct, 151
Innovative
　brand communication, 264
　thinking, 33, 43
Insolvency, 131–139, 141–143, 145, 146
　law comm, 135
Intangible sublevels, 36
Integrated
　information systems, 155
　marketing communication, 256
　networks, 42
Intelligent machine, 73, 81
Interactive communication, 260, 263, 265–267
Interconnectivity, 32
Inter-construct correlations, 23
Intercultural
　communication, 44
　companies, 49
Interdependent global economy, 134
Interest-free business, 9, 11
Inter-governmental bodies, 134
International
　advertising, 208, 212, 214–221
　business, 31, 33–35, 40, 46, 51, 71, 187, 199, 203, 209, 212, 213, 220–222, 225
　　context, 34
　　deals, 51
　　organizations, 35
　　platform, 187
　commercial conciliation, 134
　communities, 51
　data corporation (IDC), 164
　human resource management (IRHM), 41, 46
　journals, 1

　law, 43
　limitations, 216
　market, 40, 51, 97, 132, 206, 207, 209, 211–213, 220, 221, 248
　　segmentation, 209
　marketing, 205, 209–213, 215–217, 219, 221, 222
　　benefits, 217
　　challenges, 219
　　characteristics, 216
　　in globalized era, 216
　organizations, 46
　promoting, 213–216, 218, 221
　segment level, 209
　showcasing, 213–221
　standards, 145
　trade, 33, 43, 134, 208, 209
　　practices, 33
Internationalization, 205, 212, 213, 222
Internet
　banking, 126, 173, 177, 178
　connected services, 17
　of things (IoT), 13–21, 23–27, 71, 86, 195, 228, 231, 234
　　affecting factors, 15
　　conveniences, 16
　　cost, 16
　　privacy and safety, 18
　　services, 25, 26
　　status, 17
Inventory management systems, 272
Investopedia, 131
Itaka, 296, 298, 300–302
ITunes, 163

J

James Gaskin's statistical tool package, 23
Jio money, 178
Job
　market, 84
　polarization, 77
Jurisdiction, 134, 136
Justified wages, 9

K

Kaiser-Mayer-Olkin (KMO), 104, 105, 111, 115

Key
  cross-cultural barriers, 43
    ethnocentrism, 44
    parochialism, 44
  customers, 145
  negotiators, 43
  professional parameters, 43
Kurtosis, 20, 24, 103, 104, 110

L

Large-scale activity, 216
Legislative guides, 134
Lewis model, 40, 41
Liberalization, 42, 192
Likert scale, 20, 103
Linear
  active, 40
  regression, 59, 62–66
LinkedIn, 164, 259, 282
Liquid crystal display circuits, 196
Liquidation, 135, 139
Liquidity crunch, 131
Literature classification, 286
Loan grantors, 132
Local showcasing, 211
Logistics and controlled atmospheric technology, 196
Long term
  benefits, 155
  goals, 38, 273
  orientation, 37, 38
    vs. short-term orientation (LTO), 37, 38
  solutions, 156
  strategic objectives, 155, 156
Low
  power index, 38
  uncertainty, 38
Loyalty, 49, 100, 102, 119–121, 123–128, 207, 254, 282, 284, 285, 288, 294
Luxury
  brand, 95–98, 100–103, 106, 112
  products, 95, 96, 104, 107, 111, 112

M

Machine
  languages, 248
  learning, 77, 78, 81, 83, 86, 164
Mandatory internet access, 184

Marketing, 3, 15, 17, 51, 77, 86, 95–102, 106, 107, 120, 126, 165, 192, 205, 206, 208–212, 218, 221, 222, 236, 253–258, 260, 265–269, 272–274, 275, 277, 281–290, 293–295, 304
  research, 256
  strategy, 51, 210, 256, 257, 273, 283, 286, 290, 304
Marlbordo, 210
Masculinity, 38
  society, 38
  vs. femininity (MAS), 37, 38
Mass
  availability, 267
  media platform, 99
Massive global competition, 13
MasterCard, 178
Materialism, 98, 99, 102, 106, 107, 110, 112, 114
Meaningful interaction, 254
Medium
  scale predictor, 62
  skilled professionals, 79
Member state, 139, 140
Metrics, 151, 152, 169
Microblogs, 101, 286
Millennials, 281, 283–290
Mining, 226
Ministry of sports and tourism (MSiT), 295
MobiKwik, 182, 183
Mobile
  banking, 14, 173, 177, 179
  commerce, 17
  payment applications, 179
Model
  fit indices, 22, 105
  insolvency law, 145, 146
  law, 132, 134–137, 141, 145
Monetary
  services, 175
  transactions, 174, 181
Morale support, 50
M-Pesa, 178
Multi-actives, 40
Multichannel banking, 119–121, 126–128
Multicultural
  business environment diversity, 42
  gaps, 43

Index 315

market environments, 33
organizations, 47
teams, 32, 42
work teams, 48
working environments, 46, 48, 52
Multidimensional
  approach, 152, 154, 169
  value systems, 33
Multi-functional managers, 42
Multinational
  companies, 48, 150, 167
  corporations, 135
  enterprises (MNE), 45, 52
  organizations, 43, 48
Multiple
  communication channels, 14
  data centers, 225
  linear regressions, 66
Multi-stage random sampling, 58
Multi-tasking, 40

N

National
  electronic fund transfer, 178
  horticulture board, 189
  payments corporation of India, 181
Neckerman, 296, 298, 301, 302
Negotiation, 35, 40, 43
Nerve center theory, 140
Neural networks, 73, 81, 83
NextGen businesses, 150, 167
Ninjacart, 192, 199–201
  achievements, 201
  operational challenges, 201
  supply chain dynamics, 201
Non-banking financial sector, 176
Non-domestic markets, 209
Non-functional associations, 102
Non-governmental bodies, 134
Non-legislative instruments, 134
Non-member states, 134
Non-work activities, 57
Null hypothesis, 114, 115

O

Oculography analysis, 305
Offline
  sector, 123
  services, 125

Omni-channel, 282
One-time password, 178
Online
  advertising, 96–100, 283, 286, 287, 289
  banking services, 17
  communication tools, 236
  communities, 284, 287
  consumers, 285, 287
  marketing, 96, 98–101
  media, 97, 106, 107
    practice, 97
  payment, 177, 178, 183
    portal, 14
  services, 123
  shopping, 17, 285
  tools, 19
  transaction, 173
Open
  ended questions, 5
  standards, 226
Openness, 57, 65–68
Organic sub-economy, 209
Organization job-changing environment, 248
Organizational
  behavior (OB), 41, 43, 46
  cultural
    integrity, 50
    model, 37
  culture, 34, 36
  development, 237
  growth, 34, 48, 51
  mission, 49
  planning, 37
  reputation, 37
Orthogonal-rotation, 104
Overlapping, 44, 154
Oxigen wallet, 179

P

Paper encyclopedia, 241
Parochialism, 33, 43, 52
Past trade relations, 134
Payment applications, 175
Paytm, 178–181, 184, 265
  mall, 179
  payments bank, 179, 180
  promotional strategies, 180

Peer-to-peer communication, 285
Personal psychology, 55, 56, 68
Personality
  characteristics, 55, 57, 59, 67, 68
  dimensions, 58, 66, 67
  traits, 56, 57
Petabytes, 202
PhonePe, 182
Physical wallet, 183
Pinterest, 96, 107, 264, 287
Plethoras, 42
Podcasts, 101
Point of sale (POS), 120, 126, 177, 179
Policymakers, 13
Policy-making, 37
Polyhouses, 191
Polytunnel technologies, 191
Porter's value chain, 153
Positive
  correlation, 56, 257
  reinforcement, 49
  relationship, 63–67
  service experience, 125
Post-globalization, 212
Post-harvest
  handling, 192
  losses, 187, 190
  mechanisms, 192
  technologies, 190
Potential
  consumers, 209, 258, 293
  misunderstandings, 44
Power distance index (PDI), 37, 38, 52
PowerPoint, 261
Pradhan Mantri Jan Dhan Yojana (PMJDY), 175
Pre-paid instruments (PPI), 174, 185
Prescriptive analytics, 275
Price sensitiveness, 16
Principal component analysis, 104
Private blockchain technology, 228
Process digitalization, 158
Product information belief, 110
Productivity, 2, 49, 50, 71, 73, 77, 79, 87, 88, 158, 192, 231, 233–236, 240–243, 246, 247, 250, 251, 270, 273–275, 277
Professional parameters, 48
Profit
  making products, 271
  maximization, 1

opportunities, 210
Proof of work (POW), 226
Psychological factors, 59

Q

QR code, 174, 179, 181
Quadratic
  effect, 59, 60
  model, 60
  relationship, 59
Qualitative
  approach, 5
  assessment, 21
Questionnaire, 19, 20, 57, 58, 102, 294

R

Radical
  BM innovation, 3
  changes, 155
Rainbow tours, 296, 298, 301–303
Ramifications, 138, 146
Real-time
  data, 266
  gross settlement, 178
  information, 17, 18
  Skype calls, 76
Reconciliation, 135
Regional healthcare companies, 76
Relationship management, 40
Residential showcasing, 212, 214, 215, 219
Responsiveness, 13, 22, 23
Risk-taking, 38
Robotics, 81
Root mean square
  error of approximation (RMSEA), 22, 103, 113, 115
  residual (RMR), 103, 115
RuPay, 178

S

Sales, 9, 10, 45, 86, 97, 111, 165, 194, 198, 209, 212, 227, 230, 253, 254, 256–259, 261, 263, 267, 268, 272, 273, 276–278, 283, 284, 288, 289
Salesforce Tencent, 164
Sampling method, 1, 19, 20
Samsung, 51
Scatter plot, 59

Second order
  business innovation dimensions, 153
  CFA, 21, 22
  reflective measurement, 22
Secondary data, 1, 5
Secure traceability, 225, 231
Self-establishing contracts, 230
Self-executing agreements, 230
Self-structured questionnaire, 19
Sensor, 14, 75, 195, 202, 203, 228, 254, 277
  resistant, 230
  technology, 195
Sensory dimension, 124, 128
Service
  focused country, 84
  sectors, 169, 249
Shareholder's agreement, 142
Short-term orientation, 38
Silver nisab, 10
Singapore law, 143
Single industrial segment, 14
Siri, 83
Skewness, 20, 24, 103, 104, 110
Skill upgradation, 88, 90
Skype, 76
Sleep disorders, 77
Smart
  objects, 15
  supply chain, 199
Social
  blogs, 101
  communication, 283
  dimension, 124, 125, 128
  media (SM), 19, 95–101, 104, 106, 107, 110–113, 115, 254, 258, 263, 264, 268, 281–290
    marketing, 95–97, 99–101, 281, 283–286, 288–290
    platforms, 98, 264, 282, 284
    promotion, 97, 98, 101
    public relations, 101
  network, 282, 287, 288
  networking, 14, 111, 282
    sites, 283
  welfare, 7
Socially beneficial niche models, 4
Societal
  ethics, 41
  norms, 91
  standards, 99

Socio-economic system, 41
Sociological factors, 56
Sociology, 41, 55
Soil factors, 192
Solitary market, 206
Stakeholder, 4, 5, 34, 45, 135, 151, 192, 225, 226, 230
  value, 159
Standard
  deviation, 62–65
  operating procedures, 49
Standardization, 210
Standardized beta coefficient, 62, 65
Starbucks, 51, 259
Startup funding, 85, 93
Stereotypes, 43, 50
Stipendiary future, 244
Structural equation modeling (SEM), 13, 20, 23, 106
Substantial wealth, 249
Supercomputer, 74
Supplementary digital transactions, 177
Supply chain, 4, 159, 163, 164, 187, 190–194, 196, 197, 199, 201–203, 225–228, 230, 231, 237
  management, 237
  modernization, 193
Supreme court, 144
Sustainability upgradation, 4
Sustainable
  business, 1, 2, 4, 9–11, 152, 154, 166
    development, 9, 10
    models, 4, 5
    society, 11
  development, 1, 3
  differentiation, 120
  entrepreneurship, 4
  growth, 3
  innovation, 3, 155
  model, 11
System-level innovation, 4

T

Tax mobilization, 6
Taxation, 1, 6, 10, 11, 46
Technological
  environment, 89, 240, 242, 273
  tools, 50

Technology acceptance model (TAM), 17, 27
Techno-savvy customer, 127, 128
Telecommuter jobs, 236
Telemanagement (TM), 153
Ten item personal inventory, 55
Tez, 182
Timestamp, 225
Top-down approach, 73, 74
Topography, 46, 81
Tourism
  agencies, 294
  industry, 124, 293, 304
Tourist supply, 305
Traditional
  banking mode, 128
  distribution channels, 120
  hard copy questionnaire, 58
  non-electronic mode, 125
Transparent paradigm structure, 47
Trivial issues, 45
TUI Poland, 296, 298, 301, 303, 304
Twitter, 96, 98, 107, 164, 282, 284, 286, 288, 289

U

Ultra-touchy nature, 216
Uncertainty avoidance (UAI), 37, 38, 52
Unhealthy business, 8, 9
Unified payments Interface (UPI), 177, 178, 180–183
Unique
  addressing schemes, 14
  selling proposition (USP), 260
United Nations Commission on International Trade Law (UNCITRAL), 132, 134–138, 141, 143, 145, 146
Universal
  advertising, 205, 206, 211, 216, 218
  credit associations, 135
  showcasing, 206, 211, 218
Unstructured supplementary service data (USSD), 177, 178
User-oriented applications, 18

V

Value,
  adding, 163
  chain, 151, 157–159, 163–165, 192, 228
  dimension, 153

innovation, digital, and enterprise (VIDE), 153, 167–169
  producing opportunities, 165
Varimax, 104
Verbal and non-verbal interactions, 47
Vertical
  coherence, 4
  value chains, 163–165
Virgos Schmit report, 136
Virtual
  advertisement expenditure, 96
  environment, 124
  presentations, 101, 259
  reality (VR), 254, 255, 266–268, 275, 277–279
  implementations, 278, 279
  teams, 43, 46
Virtualization, 242
Visualization, 266
Viztar agritech, 199
Voice inputs, 83
Volatile markets, 33
VRoom, 278

W

Wage payment, 235
Wallet applications, 178
Wealth and welfare business model, 8, 11
Web apps, 276
Weblogs, 101
Welfare
  business, 1, 6–11
  businessman, 9
Wezyr holidays, 296, 298, 301, 303–305
WhatsApp, 58, 241, 263, 282
  questionnaire, 58
Wikipedia, 107, 241, 282
Workforce inclination, 247
Work-life
  balance (WLB), 42, 55–68
  conflict, 63
  facilitation, 63
Workplace, 71–73, 77, 78, 236, 242, 244–247, 251, 276, 278
World
  bank, 132, 137, 176
  wide web (WWW), 14, 27, 282, 290

X

X-ray, 82

Y

YouTube, 96, 282, 288, 289

Z

Zakat, 7, 9, 10
Zetta, 141, 142, 144, 146
   jet Singapore, 142